IN THE WEEDS

For Claudia and Zoe

IN THE WEEDS

NURTURING COMMUNITY THROUGH LANDCARE

SONIA GRAHAM

CSIRO PUBLISHING

A catalogue record for this book is available from the National Library of Australia

ISBN: 9781486320103 (pbk)
ISBN: 9781486320110 (epdf)
ISBN: 9781486320127 (epub)

How to cite:
Graham S (2026) *In the Weeds: Nurturing Community through Landcare*. CSIRO Publishing, Melbourne.

Published by:

CSIRO Publishing
36 Gardiner Road, Clayton VIC 3168
Private Bag 10, Clayton South VIC 3169
Australia

Telephone: [+613] 9545 8555
Email: csiropublishing@csiro.au
Website: www.publishing.csiro.au
Sign up to our email alerts: publishing.csiro.au/earlyalert

Cover illustrations by Alena Batura, Ekaterina Kim, Anna Nedelchuk and Nastya Moon Designs, via iStock.

Edited by Janet Walker
Cover design by Cath Pirret Design
Typeset by Envisage Information Technology
Index by Indexicana
Printed by Ingram Lightning Source

CSIRO acknowledges the Traditional Owners of the lands that we live and work on across Australia and pays its respect to Elders past and present. CSIRO recognises that Aboriginal and Torres Strait Islander peoples have made and will continue to make extraordinary contributions to all aspects of Australian life including culture, economy and science. The use of Western science in this publication should not be interpreted as diminishing the knowledge of plants, animals and environment from Indigenous ecological knowledge systems.

Jun26_RP_ILS

FOREWORD

In the Weeds is an important and timely publication and I hope lots of people get to read it.

Why? Because Australia is under siege from the scourge of weeds; strangling native trees and smothering native grasses, obliterating paddocks, roadside verges and infesting gardens. Weeds are literally everywhere – other than in some of our big national parks – and getting them under control is essential to building and maintaining healthy, resilient and productive landscapes.

The big threat weeds pose to our environment is not always well understood but – attention federal and state treasurers – they are costing us a bomb.

In a recently urbanised and farmed country it's easy to assume that the landscape has always looked like it does now. Yet below the surface over the past 200 years there have been substantial changes in vegetation. The kinds of trees and plants, some that dominate great swathes of the country, have changed dramatically, and we are witness to an escalating decline in the ecological health of land and river systems Australia wide in part due to the attack by weeds on many fronts.

The proliferation of weeds just about everywhere, often a by-product of excessive land clearing and a habitual deafness to the caring for country wisdom of First Peoples, presents a direct challenge to the ecology and economy of the natural world.

Weeds are the least likely candidate to end up front page and front of mind, but given the immense costs they impose on nature and productive landscapes they should be in the spotlight, as should the community Landcare groups and volunteers that work hard at bringing them under control.

Sonia Graham's book is timely and well researched, with in-depth case studies of local groups who are now well established and successful in making a real difference in their area, reclaiming gullies and valleys and people's properties from the assault of toxic invaders in all manner of circumstances.

Of particular importance when thinking hard about how to replicate and expand these efforts, is the insight that without a ramped up national conversation about weeds, and, in the absence of greater collaboration across sectors, farm fences and ecosystems, the blight of weeds on the march will continue to threaten the fabric of natural ecosystems, indeed of our way of life.

In the Weeds is an invaluable primer helping us better understand why and how we can greatly amplify efforts on the weed front, building community and natural well being at the same time. *In the Weeds* shows us that taking on weeds can be a win-win; increasing social capital and restoring degraded lands. So let this be a national project that brings us together – now.

Hon Peter Garrett AM FTSE

CONTENTS

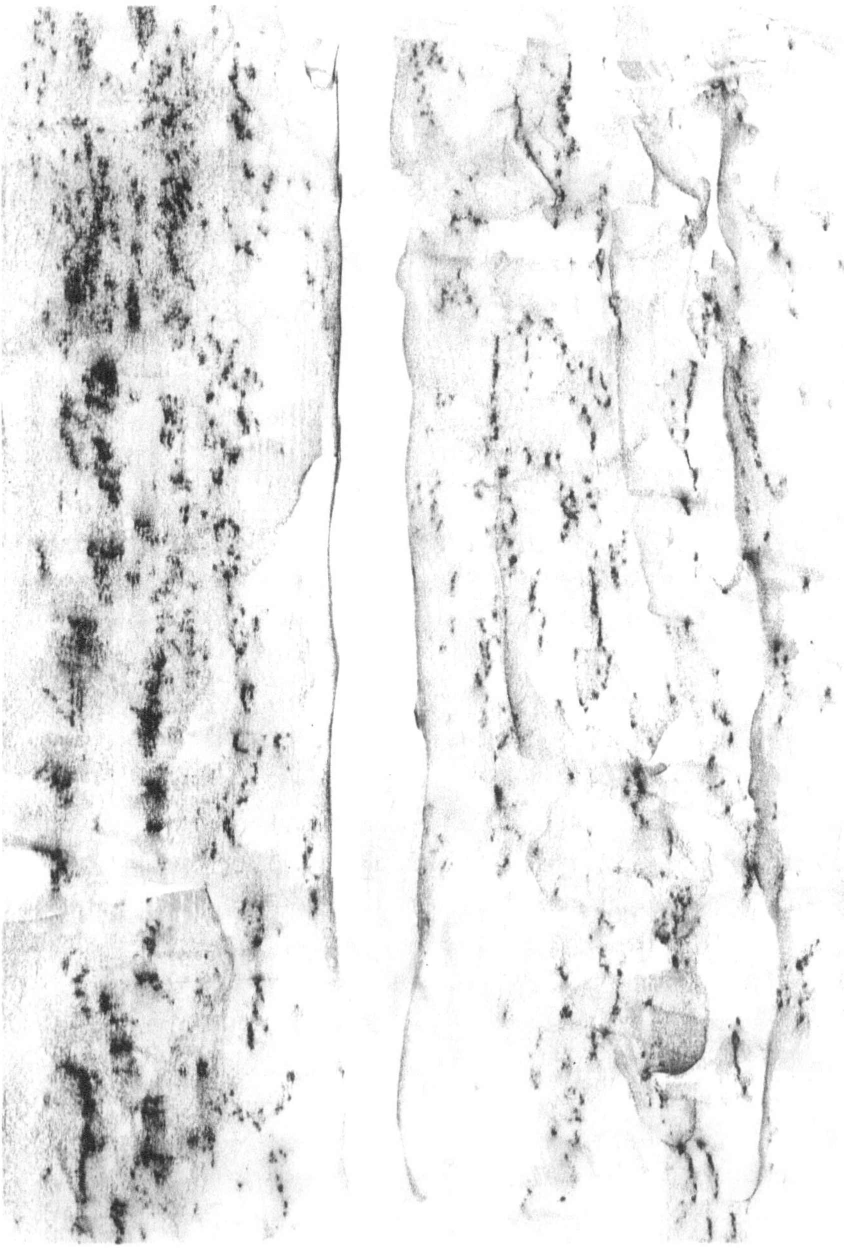

Sitting with the casuarina bark. (See Chapter 2.)

PREFACE

The inspiration for this book came from Towamba Valley Landcare. I first met the group in 2009 when I was doing my PhD. I had already conducted dozens of interviews with landholders and weed officers across central Victoria and south-eastern New South Wales (NSW), Australia. I had heard countless individual stories of heartache and desperation in the fight to control serrated tussock (*Nassella trichotoma*). The seeds kept blowing in and the weed keep thriving, reducing the value of grazing land and increasing the risk of fires. Some landholders who felt that their neighbours weren't doing enough to control serrated tussock would report them to the authorities. Others were afraid of contacting the weed officer for fear of receiving a fine for not having their weeds under control. There was very little trust and almost no one was working together.

But then I came across the Serrated Tussock Working Group of Towamba Valley Landcare. In this small valley in the far south of NSW, a group of four landholders had decided they needed a valley-wide response to serrated tussock. They realised that to minimise the impact of serrated tussock on grazing, the main source of income in the region, would require everyone doing their part. This included private landholders, public land managers – the local council, NSW National Parks and Wildlife Service, Forestry Corporation of NSW – and private forestry managers. Together they developed a common aim – not to let a single serrated tussock plant go to seed. Everyone was accountable to one another.

Twice each year they would all meet in one landholder's shed and explain what they had done to control serrated tussock on their properties. They shared details of rates of herbicide applied and the obstacles that had got in their way. The local council weed officer was part of the group. In addition to explaining what the council was doing to control serrated tussock on roadsides, they would take note of who wasn't present to inform their inspections of rural properties, supporting the efforts of the group. Koori people were trained and paid to identify and control serrated tussock and support those who needed a helping hand. At the end of the formal meeting, everyone would stay to socialise and enjoy a barbecue. Here was collective action in practice.

This made me wonder, how many more groups were there like this around NSW? What conditions make the formation of such groups possible in the shadow of government policies that target individual landholders? Do other groups operate in the same way?

In the academic literature, there is a wealth of research about how people come together to manage environmental problems that require collective action. To date, most of the attention has been on common pool resource problems, like communal pastures (made famous by Garrett Hardin's *Science* article 'Tragedy of the Commons') and water resources (studied in depth by Elinor Ostrom, and for which she became the first woman to win a Nobel Prize in Economic Sciences).

In her book *Governing the Commons: The Evolution of Institutions for Collective Action*, Elinor Ostrom identified eight 'design principles' that are commonly found in groups that work together to manage common pool resources. While I have found that many of these principles apply to the management of invasive species in general, and weeds in particular, there are aspects of managing weeds that meaningfully differ from managing common pool resources.

A key characteristic of public good problems, like weeds, is that they are rarely neatly confined, making it difficult to draw boundaries around where and from whom help is needed. This means that those working on weeds need to be adaptable, adjusting their boundaries as the weed problem grows or shrinks. What plants are considered a weed is highly variable, often contested and strongly related to place. Decisions about what weeds to work on need to be tailored to local environments and communities. Such discussion requires strong working relationships and trust.

To foster such positive relationships, collaborative weed management groups, like the Towamba Serrated Tussock Working Group, regularly organise social activities, such as family-friendly events, dinners and barbecues. Far from being an afterthought or an add-on, such events are recognised as being fundamental to the success of the groups – they are essential to giving members a chance to reflect on what they have achieved together, to socialise and plan for what they want to do next. Sharing food provides a different way of bonding, it facilitates a more relaxed atmosphere that breaks away from the shame and stigma usually associated with having weeds on one's property. It is an avenue through which the groups become something more than weeding, they become community.

Through seven case studies, this book explores questions about what drives people to work together on weeds. It considers how members sustain groups over time – most groups in this book have been active for over 20 years – and whether there are any common factors that make them successful.

This book is the product of years of my own social research as a human geographer interested in collective action. It draws heavily on my most recent research project – an Australian Research Council Discovery Early Career Researcher Award (DE200100234) – titled 'Catalysing collective action for effective weed management'.

Primarily, this book draws on more than 100 interviews, most with members of NSW-based Landcare and Rivercare groups. Landcare was launched in Victoria in 1986 as a grassroots initiative where landholders come together to address challenges associated with managing rural land. Rivercare followed about a decade later to facilitate on-ground river management. In working closely with these groups, I've come to appreciate how remarkable the Landcare model is. It has so much going for it, from a landscape perspective, from a productively perspective, from an environmental perspective and from a social cohesion perspective.

This is not an academic text. This book was written to provide inspiration for anyone working on environmental challenges that require collaboration. This may include those who want to start a community group or are looking for ideas to reinvigorate their group. It may also help those whose groups are thriving and are looking for ideas on how to showcase their successes now and into the future.

Beyond that, this book is for people who have never heard of Landcare, who may be getting to know weeds for the first time, who may be feeling frustrated or beleaguered by the scale of their weedy challenge. If you're not sure about weeds, or what to do about them, or simply want to get to know your neighbours, I hope the stories in this book give you all the reasons you need to start a conversation about weeds and feel socially connected to those around you.

1

Introduction

Weeds are all around us. We pass them on every road we drive, every train we catch and every path we walk. It doesn't matter if we're in a big city, out in a paddock or even a national park, there is bound to be a weed nearby. Yet despite their prevalence, many of us are blind to weeds. They can blend into the greenery of road verges, the river corridors and fence lines.

Even places that might be thought to be relatively 'pristine' have not escaped the spread of weeds.

The iconic Kosciuszko National Park, in south-east Australia, is currently in the throes of one of the best-funded weed detection and eradication programmes ever undertaken in Australia.[1] Millions of dollars and thousands of hours of paid and volunteer labour as well as highly trained detector dogs[2] have been used to detect and remove orange hawkweed (*Hieracium aurantiacum*) and mouse-ear hawkweed (*Hieracium pilosella*). This innocent-looking daisy from Europe poses a serious threat to our unique alpine environment and the farmland that surrounds it.

Meanwhile, in the heart of Australia, the custodians of Uluru-Kata Tjuta, Katiti-Petermann and Watarrka are extremely worried about buffel grass (*Cenchrus ciliaris* and *Cenchrus pennisetiformis*). Buffel grass not only makes Country inhospitable to native plants and animals but also increases the frequency and severity of fires.[3] They see it as an 'enormous issue … we're not seeing the many animal and plant species that are significant to us because so many that would normally be in our Country have been choked out [by] this awful grass [that] has spread everywhere.'[4] There is so little government funding for controlling buffel grass that Rangers work in their free time to remove it and give culturally significant sites a better chance to recover.

Examples like these abound across Australia. Indeed, scientists note that weeds are one of the key threats driving land degradation and the extinction of native species.[5]

Beyond environmental and cultural impacts, weeds cost the Australian economy billions of dollars each year in lost agricultural production and in the cost of applying weed control. Most crops and pastures have weeds. Farmers are constantly vigilant; many take a spray unit with them as they go about their work to control any weeds they find along the way. On average, Australian farming families spend one month each year controlling weeds and a similar amount of time is spent by contractors and other staff.[6]

Weed control is also expensive. A recent survey of wheat, barley and canola farmers found that on average they spend one-quarter of their input costs managing weeds.[7] For some farmers, that figure was as high as 80 per cent. And those costs will continue to push up the prices of our everyday foods as weeds become more resistant to the world's cheapest and most widely used herbicide, Roundup® (glyphosate).

Just like our changing climate, weeds are slowly but surely having profound environmental, economic and social impacts. However, unlike climate change, there is no national conversation about weeds. In a large part this is because many of us simply don't notice the problems they're causing.

What happens when we don't talk about weeds?

There are significant ramifications of our societal silence on weeds, aside from slowing down the speed with which information spreads and missing opportunities to control weeds early. The silence means that farmers and other rural landholders often feel that they have no one who they can talk to about weeds. This is especially true if they don't want to involve or don't trust the local weed officer. With no one to turn to about weeds, the psychological impacts can be profound.

One of the farmers I talked to in 2008, explained how the spread of serrated tussock (*Nassella trichotoma*) was so concerning to them and so overwhelming that it had driven them to consider suicide.[8] Other landholders described the spread of serrated tussock as 'frightening' and 'devastating'. This sense of exasperation was only exacerbated by government policies.

The use of fines by governments to penalise landholders for not adequately controlling weeds provides an additional source of stress and anxiety, with some landholders explaining how weed control directives would 'put them over the edge'.

Beyond the direct emotional impacts on individuals, enforcement-oriented government policies discourage rural landholders from talking about weeds. Policies convey a message that having weeds is bad, so if someone has weeds, it means that they haven't been managing their land well. As a result, neighbours feel uncomfortable raising the issue with one another. Any comment they make about weeds they observe over the fence is likely to be interpreted negatively. Such policies and social norms discourage landholders from talking to other landholders and offering advice on weeds or can result in conflict when landholders decide to say something.

This phenomenon is well captured by one of the weed officers I interviewed during my PhD:

> There can be quite a bit of friction between landholders over weeds. You've got one landholder trying to do his [sic] utmost to keep his property free of weeds and you've got another landholder on the other side of the fence [who] doesn't give a damn about it. It does cause a lot of strife and that amongst landholders. Landholders always try to keep, be cordial with their neighbours because they've got other issues – they've got fences and so forth. But out of frustration they'll come to us and ask us to address it and usually with a proviso saying, 'don't tell them I came and talked to you'.[9]

More recently, in the aftermath of the 2019–2020 Black Summer bushfires in south-east Australia, rural landholders felt isolated and overwhelmed by weeds. As burnt-out properties changed from black to green again, the sense of optimism and hope that came with the green shoots were soon dashed by the realisation that what was coming back first were plants that they didn't recognise. Weed officers and bushfire recovery staff were inundated with phone calls from distressed residents wanting to know what the new plants were and what they could do to curb their growth. They were unable to keep pace with the demand for information, with landholders languishing for months without a response about what to do.

Even plants that landholders did recognise, like the native black wattle (*Acacia decurrens*), were being redefined as a weed because of the vigour with which it was growing. It struck fear into the heart of residents who saw it as the fuel for the next lot of bushfires. And so, landholders began investing significant time and resources into controlling this 'weed'. Government funds were sought to assist them in their control efforts. Knowing that black wattle is likely to thin itself out in 10 years' time provides little comfort to landholders who see the writing on the wall: bushfires are becoming more frequent and 10 years is too long to wait for nature to take its course.

Not talking about weeds not only causes an ecological and agricultural problem but it also causes a social problem. It results in landholders feeling alone and isolated. It adds to feelings of being overwhelmed after natural disasters. It causes tensions among landholders who feel that it is easier to report their neighbour to a weed officer than to have a chat over the fence. It gets in the way of building relationships with other members of the community that could provide a whole range of benefits in addition to controlling weeds.

We need to work together on weeds

Weeds move easily across property boundaries, spread by wind, floods, fires, animals and people. Because of the ease with which they spread, the last thing we should do is deal with them on our own; we need collective responses and approaches to managing weeds. We need people to talk to their neighbours and new residents about weeds of concern and their impacts, whether the weeds are feasible to control, and how that fits into caring for Country.

Residents affected by the 2019–2020 Black Summer bushfires wanted someone who they could turn to for advice on weeds. While many landholders use existing websites and apps to identify weeds, they rarely trust that the identifications are correct. They want reassurances from long-term landholders or government staff. Landholders are concerned about doing the wrong thing, such as accidentally removing native plants instead of weeds, or causing erosion by removing weeds in burnt areas. These concerns can prevent landholders from rapidly controlling new weeds, allowing weeds to form seedbanks that become long-term problems.

Beyond talking, we also need people to work together on weeds. If one person is controlling weeds but a neighbour isn't, their efforts will largely be futile. Regardless of who I have talked to about weeds, right across Australia, there is widespread agreement that we need to work together to manage weeds and that there are many benefits that would come from doing so. Farmers agree that working together would lead to greater reductions in weed populations than working on them alone, and over time would require them to spend less time and energy controlling weeds. Collaborative approaches are also seen to have benefits for the quality and quantity of agricultural yields, as well as the added benefit of getting to spend more time with local landholders.[10]

Despite widespread agreement of the benefits that would come from greater communication about weeds and greater collaboration, there are remarkably few people who work together on weeds. In some cropping areas in Australia, as few as 18 per cent of growers work with other landholders on weeds.[11] In grazing areas, few landholders report working with their neighbours on weeds.[12] Because the examples of successful collaborations are so few and far between, landholders often don't have a sense of what benefits collective weed management could provide, or where to start.

The silence on weeds and agreement about the benefits that would come from talking and working on weeds provide the motivation for this book. First, I want to spark a conversation about weeds. I want to show how weeds need to be part of our conversations about the future and about how we prepare for and respond to our changing climate. Weeds are, and have long been, changing our relationship to Country; they are thriving in our disturbed environments, responding with vigour to increased carbon emissions[13,14] and increasing likelihood of devastating fires.[15] Second, I want to shine a light on seven communities who have overcome social norms that discourage talking about weeds and found ways to work together. In doing so, I want to highlight the diverse social benefits that come when communities work together on weeds, and provide inspiration to those who would like to work with others.

Plants out of place?

To be able to work together first requires agreement on what plants are deemed weeds, which is highly subjective and often hotly debated.

Contemporary Aboriginal and Torres Strait Islander peoples' perspectives on weeds highlight the importance of understanding weeds in context; how they interact with other plants and animals, and how they enable humans to engage with one another and in cultural practices. Crystal Arnold, a Gundungurra woman from south-east NSW, discusses the importance of considering weeds as part of Country. In engaging with weeds on Yuin Country, she explains how weeds take up space at the expense of other plants, how they can provide care and shelter to others and how they contribute to how we experience places.[16] Recognising weeds as being part of Country requires us to reflect on how we can find a balance between removing weeds and accepting their likely persistence.

Thomas Bach's work with Aboriginal communities in northern Australia reveals how weed removal that is divorced from context is perceived to be frustrating, pointless and a waste of time.[17] From this perspective, plants need to be evaluated as to whether they 'belong' in Country. This includes considering whether the plants are edible, provide habitat or fodder, and whether they help to restore degraded landscapes. Elders from Bardi Jawi, Bunuba, Nyikina Mangala, Ngurrara and Wilinggin Country in the Kimberley of Western Australia felt weed work was most important when it enhanced knowledge of and connections to Country and made Country healthier.[18]

Aboriginal and Torres Strait Islander peoples' approaches to managing weeds stand in stark contrast to colonial approaches to managing weeds, which have long sought to be much more absolute in declaring certain plants weeds, regardless of context. Starting with the 1851 *Scotch Thistle Act* in South Australia, Scotch thistles (*Onopordum acanthium*) were declared to be plants in need of destruction across the province. Any landholders who were found to have Scotch thistle were given 7 days to remove all weeds from their private property. If they failed to do so, they could be fined between 2 and 10 pounds. This punitive approach to weed management laid the groundwork for how weeds were to be managed in Australia for the next 170 years.

To this day, plants that are considered problematic in a local government area, region, state or territory are added to noxious or priority weeds lists. Landholders are required to control such weeds, with penalties in place for those who do not (although there has been some shift away from the use of fines in some states over the last decade). Implicit in all weed-related Acts is a recognition that weeds are mobile and require everyone to do their part to control their spread. Yet the focus has largely been on ensuring that individuals are managing weeds rather than working together across a landscape.

How we define and perceive certain species of plants as weeds changes over time and place.

In the early 1900s, debates about whether specific plant species were weeds played out in newspapers. Minutes of local council meetings recorded the pros and cons of what are now considered to be 'widespread' weeds. For example, in Victoria there were local councillors who saw African boxthorn (*Lycium ferocissimum* Miers) as useful for creating hedges. Other councillors who saw it as habitat for pest animals wanted the plant declared a 'noxious weed', which would require all landholders in the district to control it. Once declared noxious, the same newspapers reported which landholders were taken to court for not adequately controlling their weeds, and how much they were required to pay in fines. People were named and shamed. As the punitive approach gained momentum, social norms evolved that discouraged landholders from approaching one another and talking about weeds.[19]

Today, each state and territory in Australia has its own lists of noxious weeds and associated Acts that specify the ramifications of landholders not managing those weeds. In NSW, local governments have their own priority weeds, in recognition that weeds are more problematic in some areas than others and require different levels of investment in identification and control. The 1993 *Noxious Weeds Act* in NSW was very similar to the 1851 South Australian *Scotch Thistle Act*. Landholders could be fined and prosecuted for not adequately controlling weeds that were declared to be noxious in the local area. Indeed, in Cooma, NSW, one famous case resulted in a landholder having to pay more than $100,000 for the weed control costs that the local council incurred from spraying weeds on their property.[20] Over time, the weeds that were deemed a priority changed, but the government approach to engaging with 'recalcitrant' landholders had not.

In 1999, the federal government developed a national list of the most problematic weeds: the Weeds of National Significance.[21,22] Initially, a list of 20 weeds were scientifically ranked and declared to be the worst weeds. In 2012, a further 12 weeds were added to the list and in 2024 nominations opened for a third tranche of weeds to be added.[23] Here again, we see a desire to specify particular plants as problematic, regardless of context. Far from being locally determined, these lists attempt to impose order on both the plants and the people on whose land they exist. The often-presumed neutrality of government-mandated list-based approaches do not effectively account for the myriad ways people perceive and manage weeds. Instead, there is a range of social, aesthetic and other value-laden reasons behind how people decide what plants are weeds and how to deal with them.

Just as Aboriginal and Torres Strait Islander peoples may feel frustrated by having to control weeds that governments deem to be a 'priority' but don't match with cultural concerns and practices, many rural communities feel that government priorities are at odds with their own. In the Darling Downs region of Queensland, croppers are particularly concerned about Feathertop Rhodes grass (*Chloris virgata* Sw.), a weed that is difficult to control using herbicides and that is becoming more widespread.[24] Yet, their concerns are not reflected in the state's priority weeds list which determines the weeds local councils must control. So Feathertop Rhodes grass flourishes on roadsides, which are treated by councils for other weeds. This only infuriates local landholders in Queensland and elsewhere who believe their efforts to control this weed are undermined by the lack of effort by public land managers.[25,26]

This brings us back to the question of what a weed is and how we can agree on which weeds to prioritise. Successive local, regional, state, territory and federal government organisations have attempted to identify and define which plants are most worrying to their constituents, based on their impacts on culturally significant sites, environmentally important ecosystems and to agriculture. As the chapters in this book will show, the lists often lag behind public concern or do not focus specifically enough on the cultural, environmental or agricultural assets that the public wishes to protect. This means there is often a mismatch between government support for controlling weeds and community concerns.

More recently, governments have turned their attention to 'new and emerging weeds' because managing these plants is perceived to have the greatest return on investment with regards to minimising their potential impacts. The challenge with this approach is that the weeds listed often run ahead of community concern, and if a community does not perceive something to be a weed, then it can be difficult to get that community to focus on it. Understanding weeds, the policies that govern their management and how people respond requires an approach that is sensitive to the nuances of place – the physical and social environment. This is where Landcare comes in.

Nurturing community through Landcare

All the local groups discussed in this book are Landcare or Rivercare groups. Landcare was formally launched in 1986[27] and is often heralded as an Australian success story. It is a nation-wide 'grassroots' initiative where landholders come together to address multiple environmental challenges associated with managing rural land in Australia. This includes managing soil erosion and salinity, planting trees and protecting native species, and managing pest plants and animals.

Few people outside the movement are aware of exactly what it means to be part of a Landcare group. In part this is because it is only possible to catch glimpses of Landcare groups from the Internet, with a smattering of descriptions of project activities. It is also difficult to know what it means to be part of a Landcare group because each group is different.

The frequency with which groups meet, the way they are organised, what their goals are, whether they seek funding and who they get their funding from (whether it is local, regional, state, territory or federal government funding, or funding from the private sector and non-government organisations) are all highly variable. Landcare groups also take many forms, including Beachcare, Bushcare, Coastcare, Dunecare, Rivercare, and 'Friends of' groups, which often focus on public land and span both urban and rural areas. In this book, I refer to all these groups as Landcare groups, with the exception of Deua Rivercare.

Some might say that Landcare's heyday was in the 1990s when there was ample government funding to support the programme and when membership numbers were growing exponentially. It is estimated that in 1985 there were only 100 Landcare groups across Australia,[28] by 2000 there were ~5,000 groups.[29] Today there are an estimated 6,000 groups.[30] While the number of groups has largely plateaued, Landcare is still alive and thriving in many communities around Australia.

Beyond local groups, there is a hierarchy of supporting Landcare organisations. These include district umbrella Landcare networks that in NSW more or less map onto local government area boundaries, regional Landcare organisations that map onto catchment boundaries, state-based Landcare organisations and two national-level bodies – the National Landcare Network and Landcare Australia, each with their own functions. Many Landcare groups have little engagement with the umbrella organisations. Often just one or two people from each group will engage with the district Landcare organisation; many members have little to no knowledge about the broader organisational structure. For this reason, this book does not refer much to the district, regional, state or national Landcare organisations. What is important to know is that there are Local Landcare coordinators who are available to support groups, often helping to obtain or manage grants, and enabling knowledge to be shared among Landcare groups.

As with any social movement or collective action initiative, the persistence of Landcare hasn't been without its challenges. By the late 1990s, academics were concerned about the future of the Landcare movement and whether it would continue to receive the government funding it needed.[31,32,33,34] While the movement has endured, many groups have waxed and waned over the years. Keeping groups going requires active recruitment of new members and adjusting when longer-term members and leaders leave. To keep Landcare going, this book calls younger generations to action and encourages governments to continue investing in this unique organisation.

Weeding the length and breadth of the South Coast of New South Wales

The seven case studies are all located within Yuin, Wodi Wodi and Dharawal Country (Fig. 1.1). They span ~300 km, the length of the South Coast of NSW, and reach up towards the escarpment of the Great Dividing Range. To get to the groups I mostly drive down the Princes Highway from Sydney to the South Coast; I've done it so many times that I feel like it's a well-worn path, with each trip building a sense of familiarity of the twists and turns, the townships, rocks and native vegetation along the way.

There have been massive changes to the vegetation since 2019. This project started while the 2019–2020 Black Summer bushfires were very much ablaze. The smoke along the coast and inland made it difficult to breathe, especially with P2 (N95) masks having been sold out at every Bunnings hardware store on the south-eastern seaboard of Australia. All the communities I've worked with were impacted by bushfires, either the Black Summer bushfires or the 2018 bushfire in Tathra, NSW. Most of the groups were also heavily impacted by the three consecutive La Niña years that followed. Half of the groups experienced weeks or months of isolation as a result of landslides and road closures. Some have experienced repeated deluges or floods that have only added new layers of change to the vegetation.

After the fires, my first trips down the coast told a story of patchiness. In some places, where the fires burnt really hot, blackened trees are still solemnly standing. Other places were untouched, unscathed, the vegetation living to see another year, acting as havens for many native and pest animals. In between there were places that had been burnt, but not so ferociously. The eucalypts were rapidly resprouting their 'woolly

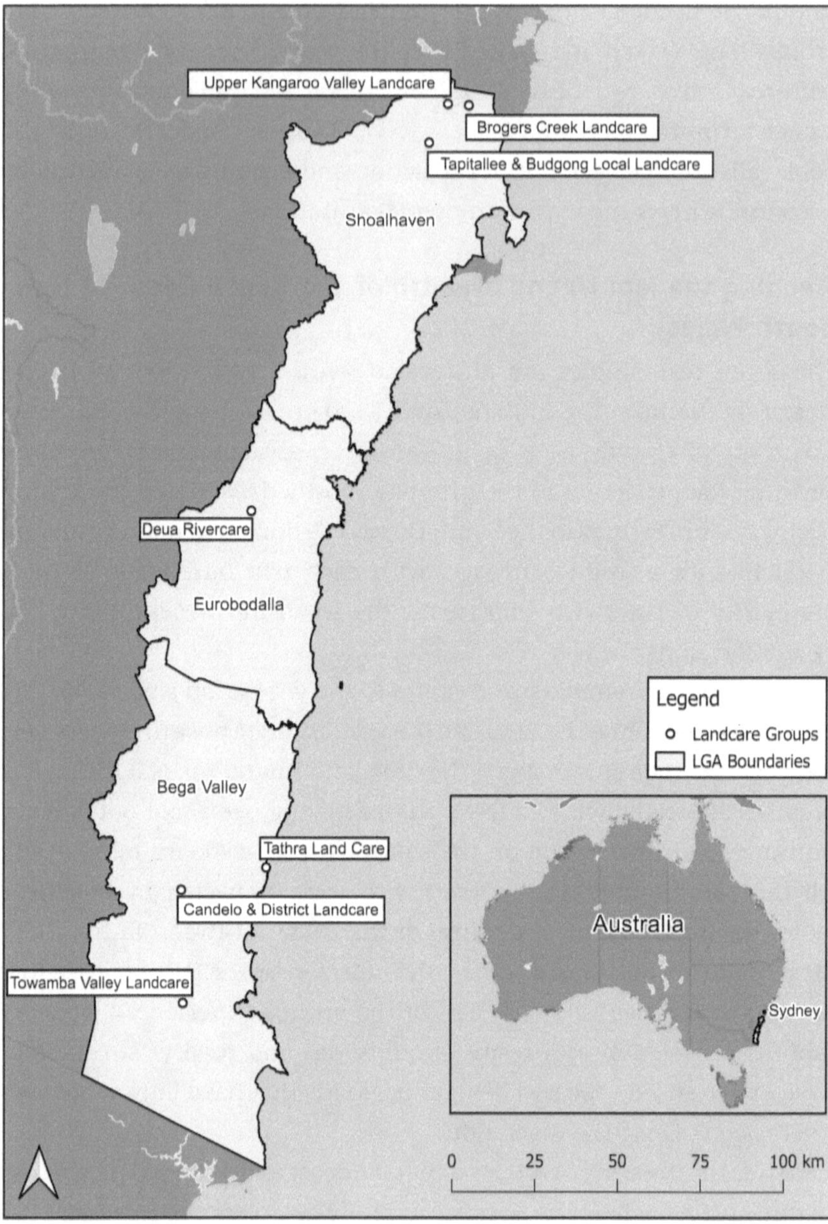

Fig. 1.1. Map of seven Landcare groups on the South Coast of NSW, Australia, produced by Teaniel Mifsud.

coats', as one of my ecologist colleagues described them. And over time, the wattles have formed a dense thicket along the burnt roadsides. Now it is getting harder to see where the bushfires were.

In amongst the re-emerging native plants were the weeds, some brought in through the strong weather associated with the bushfires, others finding a new lease on life after being burned and finding there was little competition to keep them out.

The recent experiences of communities with bushfires and COVID-19 are an important part of the story of the groups in this book. The bushfires and floods wreaked havoc on the lives and local environments of the Landcare groups. Some had to rebuild homes, sheds and other key infrastructure all while dealing with COVID-19 restrictions that made meetings, even outdoor working bees, illegal. The regular meeting patterns were interrupted; requirements about the need for vaccinations further pulled at the social fabric of the groups. While the relationships that had been built up over decades of Landcare and Rivercare events helped to provide a source of connection, they also suffered from not being tended to regularly.

As the restrictions eased, it has taken years for many of the groups to reach pre-COVID-19 levels of attendance. Some have come to the brink of folding, with emergency meetings held to determine whether there was enough interest, passion and energy to continue. In this way, the groups described here are also windows into the fate of volunteer groups across NSW and Australia. For decades, the numbers of people volunteering have been in decline, but with the COVID-19 restrictions limiting movement and social interactions many community groups have been struggling to rebuild momentum. This is something that groups like Volunteering Australia are grappling with.[35] For now, the seven groups continue to exist, but not all operate in the same way as they did before the bushfires, COVID-19 and the floods.

Gathering social data

This book draws on data collected as part of a research project that began in 2020. The seven groups selected to feature in this book are the product of long-term relationships as well as serendipity. In late 2021, I put a call out through Local Landcare coordinators and local government weed officers from across south-east NSW – from the Snowy Mountains and Tablelands to the coast – for groups who might be interested in being involved in this research, who were happy to share their stories and for me to observe and participate in group activities.

There were some groups I had studied previously who I really wanted the opportunity to engage with again, like the Towamba Serrated Tussock Working Group, to see how they had fared in the intervening period. Some I had heard about through my networks, like the Deua Rivercare group, and hoped they would volunteer. Others, like Brogers Creek Landcare, have had relationships with some of my colleagues at the University of Wollongong, NSW, and generously give their time to researchers. The only requirement on my part was that the groups were primarily concerned with managing weeds as a community.

The result of this self-selecting approach is that all the groups are located within 100 km of the coast, six of the groups primarily focus on environmental weeds and one focuses on agricultural weeds. Across the groups, members use their rural properties for primary production or lifestyle reasons. What I hope this book shows is that regardless of where groups are located and whether the purpose of the groups is environmental or agricultural, there are broader lessons about how to work together on weeds.

When I originally proposed this project, I anticipated an orderly data collection process. I would visit each group once and interview members of the group, residents of the broader community and government staff who engaged with each group around the time of visiting the group. This plan quickly went out the window. Visiting a group once is useful, but it takes repeated visits to build trust, to really get a sense of how a group operates, who the regular members are and how members relate to one another. Given the nature of the disasters each group faced, making time for interviews was also far from straightforward. What unfolded was fieldwork that spanned three-and-a-half years.

The working bees for each group are marked in my calendar months in advance. For most groups, I have been to at least three working bees, some as many as six, others only one.

For each group I have interviewed at least seven members, as well as at least two residents from the community who have decided not to join the groups, to provide broader perspectives. I have also interviewed at least three paid staff from local government, state government, research organisations or Landcare to understand what it is like for them to engage with these groups.

The descriptions in the chapters that follow come from over 100 interviews and participation in meetings and working bees. I have included quotations from the interviews to help give a sense of each group in the words of Landcare and community members. I have also supplemented the descriptions with documentary evidence. Some groups have maintained meticulous records of where they have done work, who they have applied to grants for and the outcomes of those grants. Some have consistently published articles in the local newspapers, documenting group efforts and calling others to action. Some have produced pamphlets and books. I have drawn on these additional sources, as well as academic material, to provide context for how and why these groups came about and the challenges they've faced.

What makes the groups work?

As the chapters in this book make clear, there are many ways that the groups are similar and could be grouped together. Geographically, the three northernmost groups are located in and around Kangaroo Valley, NSW, are contained within the borders of Shoalhaven City Council and are all part of the Shoalhaven Landcare umbrella organisation. Although these groups are located within 20 km of one another, as the kookaburra flies, they focus on different weeds. Upper Kangaroo Valley Landcare (Chapter 4) is mostly concerned with moth vine (*Araujia sericifera*), Madeira vine (*Anredera cordifolia*) and wild tobacco (*Solanum mauritianum*) on public land. Brogers Creek Landcare (Chapter 3) is primarily concerned with small-leaf privet (*Ligustrum sinense*), lantana (*Lantana camara*) and wild tobacco on private and public property. Tapitallee Budgong Local Landcare (Chapter 5) mainly focuses on lantana on private properties, although recently their attention has turned to black wattle.

The three southernmost groups are located within Bega Valley Shire Council and come under the umbrella of the Far South Coast Landcare Association, but are very different with respect to the weeds of concern and where they work. Tathra Land Care (Chapter 7) almost exclusively focuses on bitou bush (*Chrysanthemoides monilifera* ssp. *rotundata*) on public land around Tathra; it is the only one of the seven groups that is focused on the coast. Candelo and District Landcare (Chapter 8) manages weeds, originally crack willow (*Salix fragilis* var. *fragilis*) and weeping willow (*Salix babylonica*), along Candelo Creek. Towamba Valley Landcare (Chapter 9) was originally most concerned with controlling serrated tussock on both public and private land throughout the Towamba Valley, but has turned its attention to African lovegrass (*Eragrostis curvula*) more recently as well as other environmental issues, such as carbon sequestration.

Deua Rivercare (Chapter 6), located within Eurobodalla Shire Council and part of Eurobodalla Landcare, is the sole Rivercare group covered in this book, although its focus on riverine weeds makes it similar in many ways to Upper Kangaroo Valley Landcare.

Through the chapters, I reflect on the weeds that each group focuses on, the implications of working on private or public land, the frequency of meetings and how the groups socialise and engage on a range of other environmental issues. I consider the ways in which each group is unique with respect to the histories of the communities. I explore the extent to which the group's activities reflect or precede government attention on particular weeds and the extent to which they work independently or draw on government assistance.

Between the sections of the book, I have included interludes and recipes. These are used to capture and convey the social fabric of the Landcare groups. Just as members stop after a working bee to reflect on work done and to socialise, the interludes and recipes highlight the importance of the social connections that sustain these groups.

Almost all the Landcare groups in this book have existed for 20 years or more. Remarkably, many of the groups have had the same person or people leading the group this whole time, providing a stark comparison to the high rates of turnover of staff in government agencies, let alone the rate at which the government departments change names. The long-term commitment and dedication of these grassroots community groups provide consistency well beyond what any funding or political cycle can provide. In doing so, they also plug a gap in rural community services.

Where a conversation about weeds can take you

The stories in this book attest to the value of starting a conversation about weeds. The experiences of each group highlight the importance of collective action in agricultural and environmental settings. Together, the groups have reduced weed numbers, enhanced the productivity of the land and rehabilitated natural ecosystems. Yet they have all come to be so much more than that.

In each place, the Landcare groups provide a place where people come together, get to know one another, care for one another and for Country. Landcare provides a community for those who are new to rural areas and want to get to know their neighbours. Landcare holds a wealth of knowledge for those who want to learn more about their surroundings and what they can do to help. Landcare also provides a space for those who simply love the satisfaction of pulling, poisoning or chain sawing weeds.

In the face of a changing climate, of more frequent and intense bushfires and floods, Landcare also provides the social capital needed to help those affected. It is a place where those who have lost their homes to bushfires can find reprieve, and an avenue through which those in the broader community can support them.

Now, as much as ever, Landcare is an important grassroots movement that enables essential conversations to be had. It is a highly adaptable organisation that enables communities to come together in ways that suit the people, the place and the plants.

2

Caring for Yuin, Wodi Wodi and Dharawal Country

This chapter foregrounds and acknowledges the significance of Yuin, Wodi Wodi and Dharawal Country. For Aboriginal and Torres Strait Islander peoples, Country is a term with many layers of meaning that is used to describe land, water, animals, plants and people as well as the connections between them.[1,2,3] It holds complex ideas about laws, lore, knowledges, languages, customs, practices, histories, and identities.[4,5] It is more than physical elements. It embodies Ancestors, spirit, wind, rocks and self.[6] Country encapsulates how you connect to place and with others in that place.[7]

Where this research took place, Country is deeply connected to Yuin, Wodi Wodi and Dharawal people, their customs, identities and knowledge systems that have evolved over millennia. It is given voice and expressed through their many languages, stories, songs and dreams. It is much bigger than these individual elements, connecting each of us with them over time and from one place to another.

Country, in the most respectful way that I can explain as a non-Aboriginal woman, makes itself known and felt throughout this book. It is the weeds and the way they have called the seven Landcare groups to action. It is the changing of the seasons that cause the weeds to flower, set seed and spread. From my observations as someone who is still learning, Country is the native plants that get crowded out or grown over by the weeds. It constitutes the bushfires and floods that have transformed the landscapes that the Landcare members live, work and play in. It changes us and is changed by us. We have learnt Country cannot be disentangled from the work that Landcare groups do, nor the relationships they have built. When I use the term 'we' in this chapter, I use it to include all of us: Country, those who are Aboriginal and non-Aboriginal, Landcare volunteers and readers.

In discussing the work Landcare groups do, it is essential to recognise Country and the gentle caring practices that have long existed and that persist.[8] Country has been cared for and by Aboriginal people for millennia. To this day, Aboriginal people maintain connections, relations and responsibility to Country.[9] The Landcare groups that I have joined have only been caring for Country in the very recent past; all of us are migrants.

In much academic writing, Country is often ignored or misunderstood.[10] This chapter represents a small step along my journey towards creating a space for Country with guidance in my research, of learning through Aboriginal experiences and knowledges, and Yuin understandings of Country specifically. To listen to Country, I have had to disrupt the timelines and temporalities of academia. I have made time to engage with Aboriginal peoples, their yarnings and writings about Country. I have made time to sit with Country, to listen to Country and begin to make sense of myself in relation to Country. I hope this chapter may help encourage others to begin or further their own journey with Country.

Here I disrupt the flow of this book to share a poem that expresses my journey and evolving relationships with Country.

Cyclone Alfred may be a thousand kilometres away
But we can feel its presence
The humidity has been pushed down the coast
The rain has come and gone, come and gone
Less here than up the coast
But it reminds us of the connections between places

I spent the weekend on Yuin, Wodi Wodi and Dharawal Country
The rain brought out the leeches, slugs and snails
Slugs of a colour I've never seen before
A translucent yellow with a line of red
It reminds me of all the living beings that I am ignorant of
The changes in my environment that have passed me by
Of the lessons I haven't learned

What is the role of these slugs here?
Trained as a scientist I look to the Internet for answers

But it tells me little
A description of what they look like
Confirm that they have previously been sighted here
But not how the slug relates to this place

This slug reminds me of the limits of scientific knowledge
That there are other ways of knowing and understanding
It is incumbent on us to learn about where we are
Country will teach us if we take the time to look, listen, see
To pay attention to the changes around us
And try to make sense of what they tell us

It's a slow process
Not one that can happen in a moment
In a week, in a year
I have learned that to understand and appreciate this place
I need to take the time to come back day after day
Year after year

If we watch and hear
Country will teach us
The reasons why this slug is here
How it relates to the weather, the plants and animals around it
Why a cyclone a thousand kilometres away brought it out today
For me to cross its path

It's the same for weeds
I know that I need to look, listen and see
Watch as they move across the landscape
When they flower
What other plants and animals they befriend
Their positives and negatives

This book is just part of this journey for me
Of slowing down to look, listen and see with Country
Of trying to understand how the plants and animals
Relate to one another and where I fit
I want to unlearn what I have been taught
And relearn a different way

Yuin Country as teacher

Within Aboriginal knowledge systems, Country is teacher.[11] It can send messages, feel and take action.[12] It can be listened and talked to.[13,14] Country provides an alternative way of learning about and understanding the world. Country encourages us to go beyond thinking and acting, to feeling and being open to a spiritual understanding of our many relationships.

If we listen to it, we can understand better the ongoing effects that colonial systems have had and continue to impose. Country challenges colonial conceptualisations that separate people, animals, plants and places as well as colonial practices that seek to dominate and control Country, rather than work with it.

If an individual has an open mind to see all the living unseen and seen teachers of Country – trees, mountains, water, animals, birds, people, rocks, humidity and all the entities that make up Country – Country provides the stories and the knowledge. Country also provides the space to learn, reflect, challenge, cry, laugh, oppose and agree, while experiencing an ancient story of connection.[15]

The Landcare groups in this book all work with Yuin, Wodi Wodi and Dharawal Country. According to Yuin culture, Country is knowledge holder and teacher.[16] We are inseparable from it. To learn from, with and about Country we need to take the time to stop and observe. Be prepared to listen to the lessons it wants to share and respect that there may be stories that are not for us to hear.

During this research, I made time to work with Country alongside the Landcare groups. Over the course of countless field trips, I worked among dune vegetation, dodging spider webs and keeping a lookout for wombat holes. I was drawn into riverbeds, where frogs croaked, and corellas shrieked overhead. The height of the river changing with each visit. I worked my way up steep hills and down again. Looking for weeds, trying to distinguish them from the surrounding vegetation. I felt fortunate to be able to experience so many places and begin a journey of making time to slow down to listen to, feel, experience and learn from Country.

Beyond learnings that have come from working alongside Landcare groups, I have made time to sit alone in the places where they work to see if I could hear and feel what Country wanted to tell me.

From sitting with Country, I began to understand the world around me in a new way. I noticed how creeks flowed and shifted, leaving sandy beds that enticed new plants to grow. I watched where seedlings that were intentionally planted have thrived and others alongside them have withered or been swept away. I began to realise how during my life I have seldom gone back to the same places and observed them over time. I usually see the big picture rather than the small details.

I observed Country quietly recover from bushfires. In some places the black scars quickly turned green again, but in others the black scars remained much longer, a response to how hot the fire burned.

I have started to pay attention to where specific plants like to grow. I have noticed the affinity casuarinas have for riparian zones and that they often hang out with paperbarks (*melaleucas*). Once I started paying attention to them, they were no longer a backdrop to my activities but central to them. Casuarinas seem content in so many different places – along freshwater creeks, in parks and close to beaches. They even pop up in my garden beds.

Sometimes Country seemed abuzz with things it wanted to share. Other times it was quiet. I often had questions turning over in my mind. It seemed like Country had more to share than simply observations about the shape and form of the world around me, but I didn't have a way to understand them. For this I need cultural teachers.

Cultural teachers

Cultural teachers can help us to understand and interpret the lessons Country offers.[17] During the period of my research, I was fortunate that my Aboriginal colleagues at the University of Wollongong, Anthony McKnight and Crystal Arnold, organised community yarns with Country for my geography colleagues and me. Together we would meet on the lawn outside our building. We would sit under the shade of the big gum tree, yarning about Yuin ways of learning and being together. Anthony taught us about Yuin customs and the importance of looking, listening and seeing Country. The importance of respecting process, of having patience.

Anthony's published work talks about Mingadhuga Mingayung, which is a way in which non-Aboriginal people can be taught by Country. It involves not only considering what our mind teaches us but also our emotions and spirit.[18] It requires us to examine ourselves, to consider what we believe we already know, what we have learned through western knowledge systems, and to be attentive how Country can help us understand embodied stories of connection.

When I yarned with Anthony about this chapter and the challenges I was having in understanding the lessons that Country could teach me, he invited me to revisit my relationship with the casuarina. What else might it be trying to teach me beyond its affinity for water and its ability to thrive in many places?

Together we approached a casuarina that stood tall, shading the place we had just been sitting at. We observed it. Its height, its seed pods. Its proximity to other casuarinas and other trees. We noted the long, tender, green leaves that turn brown once they fall, covering the ground beneath.

That day there was something about the bark that drew my attention. Anthony asked me what it might be trying to tell me. We talked about a few things, but none of them felt right. There was something else, but it wasn't obvious to me yet. Anthony suggested I sit with the bark for a while longer and see what it told me. He didn't mean literally, but rather asked me to give myself time to sense and feel what lessons I might learn from the bark.

Over the next two weeks, I considered the bark. I took the time to observe other casuarinas; young and old, some covered with moss, others with none. I waited and gave myself time to allow the lesson to emerge.

After a few days of sitting with the bark, I began to understand it differently. It struck me that casuarina bark is like a puzzle. The rough bark has ridges and crevices, yet all the segments fit together. Parts of the bark are more visible and knowable than others. This is evident on the charcoal rubbing of bark on page ix.[19] The rubbing is from a single casuarina tree. The gap in the middle is a deep crevice that the charcoal couldn't reach.

When I began to understand the bark as a puzzle, it changed the way I related to casuarinas. I began to appreciate the uniqueness of individual trees. I also began to see connections between the bark of casuarinas and other native trees. In some ways the casuarina bark resembles that of lilly pilly and white cedar trees. Their bark is similar in that it is rough but different in the nature of the roughness and the way their puzzle pieces fit together.

But the idea of the puzzle seemed to be about something else too. It was about more than learning about casuarinas, about other trees, and my relationship with them. The idea of the puzzle helped me understand something I was grappling with personally.

At the time I was trying to write the conclusion to this book. The idea of the puzzle helped me to look for and see connections between the Landcare groups in a different way. It felt like a flower was blooming inside my mind as I began to appreciate similarities and differences among the Landcare groups that I hadn't considered before.

From afar, the form of the Landcare groups might seem the same, but up close each group is unique in terms of how people come together and interact with Country. Like the casuarinas, some of the younger groups are still finding their way, their patterns, making sense of how the members fit together. Some of the older groups have deeper crevices, well-known ways of doing things, that can make it more difficult to change their ways. Yet their size and maturity can inspire and nurture other groups.

I was surprised to realise that listening to Country was helping me with my research. It was helping me to explain my observations and answer questions that it hadn't occurred to me to ask before.

Indigenous ways of working with weeds

Beyond my learnings from Anthony and Crystal, I have also learned from my Aboriginal research participants. They have generously shared with me some of their ways of understanding and relating to Yuin, Wodi Wodi and Dharawal Country.

To explain what I have learned from them, I wish to share a story, with permission, told to me by one of the Aboriginal people I interviewed.

They live on Yuin Country and have been working on a Landcare weeds project for many years. Here I relate what I understood to be the key messages of the interview and what I learned from them about weeds being just one part of belonging to, living with and caring for Country. They asked me not to audio record the interview, so the story that follows comes from my notes and subsequent conversations with the interviewee.

As you'll see, the interview moved forward and back through time and zoomed in and out from the national scale to the scale of individual plants and animals. It reflects the Yuin concept of oneness, that 'directs attention to the interconnectedness of all living things'.[20]

> When the interviewee was born the Australian Constitution didn't recognise Aboriginal people as part of the population. This inexplicable treatment meant that they felt society saw them as less than human. The non-Aboriginal children at school teased and taunted them. These experiences of colonisation and racism shaped who they are today and their ongoing lack of trust in the government, regardless of which party is in power.
>
> Their experiences also mean that they see very clearly the divisions in society, between those who have privileges, power and resources and those who do not. They noted that Aboriginal people are under-represented in Landcare and are not adequately included in the management of national parks. They wondered why the people in power find it so hard to engage with Aboriginal and Torres Strait Islander peoples and involve them in the work of caring for Country. They wanted to be included, to be part of the conversation, to have their knowledge and experience recognised and valued.
>
> In among the concerns about how Aboriginal people are treated, Country wove itself into the story: 'I'm blessed because I'm living in among eucalyptus trees.'
>
> The interviewee raised with me their concerns about the impacts of the bushfires that came just before the COVID-19 pandemic. Their experience of the bushfires again represented a devaluing of Aboriginal ways of knowing and relating to

the environment. Their mob used to burn land to preserve it. The tragedy of watching Country burn and flood made them question why they hadn't been listened to with regards to fires and why governments insisted on building towns on floodplains: 'why didn't you listen to my people?'

They talked about their love of and concerns for the environment, about their spiritual connections to Country and the importance of totems. They talked about how they looked out and cared for their totem animals, black ducks.

They had been involved in weed management for many years as well as other conservation activities. Being involved in such activities provided an opportunity to engage with the land, to watch it, observe it.

The break in their weeding activities because of the COVID-19 pandemic meant that they had observed a step-change in the environment in which they usually worked. The coastline had been washed away in those two years of separation, it had fallen into the roaring seas. While this change had saddened them, they also appreciated that 'Mother Nature has to do what she's got to do'.

They talked about the places of significance to them, the places that James Cook had named on his voyage up the coast but that already had cultural names before that. They wondered, what would this Country be like if Aboriginal people had maintained ownership of it all this time?

Into the interview they brought in the animals around them. They noted the currawongs that were 'hanging around' and wondered why they are all in the one area and bothering the king parrots. They likened the currawongs to mosquitos and flies. Creepy. Pesky.

They liked being able to see the results of their work on the land over the years and feeling like they were improving the condition of the beaches while also contributing to the

> community. It provided them with an opportunity to get the 'families involved, to learn about their beaches, sacred places and bush tucker ... sharing knowledge and stories'.
>
> They loved working on and caring for Country but were worried about the impacts that humans were having. They were concerned about the way Aboriginal and non-Aboriginal people alike were losing interest in the environment, that 'they're really not interested in saving the plants and animals'. They reflected on how few people are in Landcare and that many people would benefit from getting out and killing a weed. Who is going to come and help?
>
> They reflected on how the world is now and how different it is from how it used to be. They wondered what would happen if there was another pandemic. If everyone is locked down again, they wondered who would go out on the beaches and control weeds.
>
> They were passionate about continuing to work on Landcare projects until their last breath. They talked about their dreams and the way that dreams enabled them to hear messages from their elders.
>
> 'I love nature, I love humanity, plants and animals.'
>
> They invited me to come for a visit, to see the sites where their ancestors used to frequent, where their ancestors were buried, and where people now undertake recreation and leisure activities. They offered to show me the work that they have been doing to look after the land and what they have achieved.

When I took them up on this offer, together we went to see places where they had worked to remove weeds and rubbish from Country. We yarned for hours, and I learned more about their work and the things they had learned with Country as well as from those who they had worked with in Landcare. They asked that I include a photo of Country in my book to help people appreciate and connect with it (Fig. 2.1).

For me, this story highlights the importance of recognising what has come before and of listening to and caring for Country. In this and other stories I have listened to from my cultural teachers, I have come to understand better the ongoing impacts of colonialisation on Country and myself and the importance of proactively confronting and disrupting colonial ways of thinking that aim to separate, diminish, control. It is important to embrace messiness and connectivity, and give voice to alternative ways of seeing, being and feeling.

The interviewee asked many questions about how we relate to one another and to Country. There is a wish in the story for us to change how we engage with the world around us. To spend more time outdoors, engaging with Country and caring for it. To listen to the wisdom of Aboriginal people's knowledge about how we can do things differently. These learnings are key to thinking about Landcare, the role it plays in caring for Country and how it could make a difference to our relationships with one another and Country. There is much more work to do. There is much to learn.

Fig. 2.1. Looking north towards Gulaga from Cuttagee Head.

Caring for Country

In some ways, the Landcare groups in this book provide a space to listen to and care for Country. The working bees provide time for people to come together, to learn about Country, plants, animals and how Country is changing as a result of our behaviours. It provides an opportunity to reverse some of our negative impacts.

Five years ago, despite having worked on the social relationships of weed management for over a decade, I knew very little about weeds. What they look like, how they behave and relate to the plants and animals around them, how people engage with them and how weeds bring people together. With the help of Landcare volunteers I have come to learn what a moth vine (*Araujia sericifera*), Madeira vine (*Anredera cordifolia*) and wild tobacco (*Solanum mauritianum*) look like, the way that they respond to the bushfires and what vegetation they replace. How their seeds are carried by birds and the habitats they provide. People have shared with me their own journeys of discovery and entanglements with weeds, and how weeds have brought them into closer relationships with their neighbours.

I have also learned about endemic plants and animals. One Landcare volunteer took the time to show me a native tiger orchid (*Diuris sulphurea*) that was thriving in the post-bushfire landscape. Another taught me about a wattle that was endemic to the local area and how to identify bandicoot scratchings. I watched as black wattle (*Acacia mearnsii*) thrived in the post-fire environment, thickets lining fence lines and hiding the burnt soil beneath.

Many of the Landcare volunteers lament the same things as those described in the story above. They wonder where the younger generations are, why there are so few of them when there is so much work to be done. They wonder what they can do to encourage more people to join them in their endeavours.

One of the key reasons that Landcare volunteers join and keep coming to Landcare is to learn more about the world around them. They acknowledge that there are limits to their knowledge about Country and seek out others who can help them to understand what they are observing. Some turn to the Aboriginal people in their community to help. To learn about the environment and to work alongside them to weed and care for Country. A few of the groups have sought to understand more about cultural burning and how it might form part of their work with weeds.

Many of these relationships between Landcare groups, Aboriginal people and their ways of understanding the world are in their infancy, yet there are some who have been working with Aboriginal people for decades, working in partnership to manage weeds on Country in culturally sensitive ways. As the chapters of this book reveal, one of the advantages of Landcare groups is their longevity. Their commitment to care over the course of decades means that they can take the time to build respectful and reciprocal relationships with Aboriginal people and with Country.

Decolonising research: doing research differently

Throughout my two decades of being a researcher, this is the longest I have taken to complete a single project. Six years. This slower way of researching, of revisiting places time and again, over the course of years, has made me rethink the way that I do research and how I write about the places I work.

My research proposal did not recognise Country and the methods were not designed with Aboriginal ways of knowing, being or doing research in mind. I designed the project with help from my colleagues, and once successful, I simply launched into the research project, as I had always done. I invited Aboriginal people to be involved but without first setting out to establish relationships with them or with Country. I didn't allow time for them to get to know me first, before inviting them to be part of the research and seeing whether there were any benefits that the research could provide to them. I was fortunate that some were generous in working within the limits of my approach and I have done my best to be respectful of the knowledge they have shared.

For the Landcare groups, who chose to be part, I acted reciprocally. I have contributed my labour to their working bees as a way of giving back. An extra set of hands pulling out hundreds, possibly thousands, of weeds over the course of the last few years to help make a difference to the landscapes in which they work. It obviously benefited my research too, adding richness to my observations and interpretations. I have also provided them with research outputs tailored to their individual needs: a poster for Brogers Creek, a video for Upper Kangaroo Valley Landcare, an article in *The Conversation* for Deua Rivercare.[21] And let's not forget that my homemade cakes are always well received.

While there were limitations to how the project started, this book intentionally foregrounds Country, Aboriginal people, Aboriginal knowledges and ways of doing research.

I acknowledge that I am living on stolen land, on Country from which Aboriginal people were removed, their long relationship with Country forcibly broken, their knowledge intentionally ignored.

The chapters of this book also speak about my experiences of Country and how it helps me understand each group. I acknowledge that I am a non-Aboriginal woman, a child of two migrants, who is just starting the process of learning and doing research differently. Of understanding myself and my work in relation to Country.

Writing this chapter makes me feel uncomfortable because it causes me to reflect on the limits of my knowledge, not only about Country but the experiences of Aboriginal and Torres Strait Islander peoples. I am faced with a lot of unknowns. I don't like to feel vulnerable. I don't like to make mistakes. But in sharing my discomfort and vulnerability, I hope it helps others to step outside their comfort zone and engage more fully with Country and what it has to teach us.

So here, in this chapter, we invite you, the reader, to consider your relationship with Country. If you are reading this book from the comfort of your couch or bed, you may wish to move outside to bring yourself closer to Country. Make time to put down the book. To stop, and as a many of my Aboriginal colleagues asked me, please listen and see what Country is telling you.

Part 1

Shoalhaven

3

Growing community by caring for Brogers Creek

I knew I had arrived at the right place when a sign attached to the fence announced 'Landcare at Work'. Today our 'work' would be a pot-luck lunch at what was fondly referred to in the email as the 'party pavilion'. I contributed my quiche to the feast of roast chicken, homemade spring rolls, muffins and fruit platter. We soon settled in and set to work in the cosy living room.

Food is important to Brogers Creek Landcare, as is inviting other residents of the locality of Wattamolla (affectionately known as Wattamollans) into their homes, especially newcomers, and reflecting on weedy work that has been done and what there is yet to do. The laid-back style of the pot-luck lunch is also characteristic of the group. They have no formal leadership structure, no office bearers, just a small group who are in frequent contact discussing the what, how and when of future group activities. Their passion for working together to make things happen is evident in every interaction.

At the end of two hours, it was clear that time was up. There was no lingering; there seemed to be an unspoken expectation that people would know when they had stayed long enough, and it was time to give the hosts their house back.

The arrival of 'new settlers'

The rise and decline of dairy farming in Australia are part of the foundational story of Brogers Creek Landcare.

The Wodi Wodi people numbered in their hundreds[1] when British colonisers first came to 'Gun:guru:gara:',[2] now known as Kangaroo Valley. Wodi Wodi people are documented as helping the first dairy farmers move into the area in the 1840s, transporting equipment and household items. Yet by 1891, only 11 Aboriginal and Torres Strait Islander people were recorded in the NSW census as living in Kangaroo Valley.[3] Beyond the massive social upheaval associated with removal of Wodi Wodi peoples from Country, dairy farming was also accompanied by environmental change through logging of native trees – red cedars, stringy barks and turpentines – with sawmills established in the valley in the 1930s.

One of the small dairy farming settlements was Wattamolla. By 1870 it was large enough to justify the establishment of a school, which would run for 80 years. The Parents and Citizens group provided a focal point for the community, organising many a social evening.[4] For a time, there was also a small church and a cricket team.

As the dairy farms in Kangaroo Valley closed in the 1960s and 1970s, a new wave of residents, 'new settlers', moved to the valley from Sydney and Wollongong. These new residents were less likely to come from a farming background and were more likely to be interested in 'just let[ting] the forest come back' and helping 'the native bush to regenerate'. While some of them grazed sheep and cattle, 'nobody was relying on the income to survive; no-one was farming in a serious way'. A personal account in a book about the Wattamolla valley explains this change in the population and associated environmental changes:

> By the early [19]80s most of the real farmers were gone ... The 'new settlers' have been discovering paradise ever since; families looking for peace and green and community. Most of us arrived with high expectations, remarkable confidence and a patchy skill set ... As the farmers left and closer settlement

> followed the weeds multiplied, the fences struggled and the forest started to re-assert itself. Comparing aerial shots tells the story starkly; heavily lantanaed slopes, wet gullies full of cedar seedlings, blackberry and Sally wattle.[5]

As the 'new settlers' of the 1980s and 1990s became more familiar with their surrounds, they 'discovered privet ... we didn't sort of know what was happening until we started to concentrate, and when we started to look, we discovered privet just everywhere'. Small-leaf privet (*Ligustrum sinense*, also known as narrow-leaf privet among a long list of other common names) was visible from Wattamolla Road, and it was impacting the rainforest around Brogers Creek.

Yet it wasn't just that the 'new settlers' were noticing a plant that had been there all along. As the quotation above explains, the arrival of the 'new settlers' coincided with significant land use change, more intensive dairying practices and fencing of riparian areas, which were conducive to the spread and growth of small-leaf privet. The influx of new residents and the spread of environmental weeds were thus closely intertwined.

There was also a social vacuum left by the departure of many of the long-term farming families. There was 'no Catholic Church or Scouts or Brownies or left-handed knitting group'. So, despite the small size of the community – in 2021 there were 71 residents spread across 46 households[6] – there were few avenues for residents to share their concerns.

When Andrew, a previous weekender, now full-time resident, became alarmed by the rapid spread of small-leaf privet in the 1990s, he decided to organise a 'Privet Poisoners Picnic'. He asked the Local Landcare coordinator to come to the Wattamolla valley and explain to the community how best to control privet. About a dozen residents showed up one Saturday in 1997 and learnt how to cut and paint privet, which involves cutting the stems and quickly painting them with glyphosate, a technique they still use today.[7]

Around the same time that the picnic was taking place, there was increased concern about privet among the local council and the wider community. An article in the *Sydney Morning Herald* in 1995[8] declared privet (small and broad-leafed) to be top of the 10 worst weeds. An article from the *Illawarra Mercury* in 1998,[9] explained that privet is 'a major health and environmental hazard'. When it is in flower it triggers asthma and other respiratory complaints and environmentally it is problematic because it spreads quickly, can establish a dense canopy, form monocultures and replace native vegetation. The article went on to explain 'many councils in the Sydney metropolitan region had declared privet a noxious weed' and that 'Shoalhaven City Council has already applied to NSW Agriculture to have privet declared a noxious weed in that region'. There was sufficient concern about privet that it was nominated for consideration as a Weed of National Significance. Although it was not ultimately listed, its nomination speaks to broader social concern about the weed that existed when Brogers Creek Landcare established.

The decision by Brogers Creek Landcare to use 'poison' to control privet from the outset meant that some members of the community refused to join. The remaining long-term farming families were also not represented. A nearby farmer indicated that they wouldn't join Landcare in Kangaroo Valley because the focus was on environmental weeds, like privet, rather than agricultural weeds, like fireweed (*Senecio madagascariensis*). Indeed, concerns about the use of herbicides and questions about which weed to focus on still influence who chooses to join Brogers Creek Landcare today. For those new settlers who were concerned about environmental weeds in the Wattamolla valley and were comfortable with the use of small amounts of herbicide, the Privet Poisoners Picnic acted as a spark, it 'got us galvanised'.

Nurturing community via a river, a road and a weed

The residents of Wattamolla are connected by Brogers Creek, which runs the length of this sub-catchment of Kangaroo Valley. The significance of the creek is encapsulated in the place name. A sign marking the locality of Wattamolla explains that it is an Aboriginal word that means 'clear water'. A book written about the area explains that 'wa:tta::molla' means 'place near running water'.[10] Many of the residents feel a sense of responsibility and custodianship to the river and relish swimming in it. As one member of the Landcare group explained, 'caring for Brogers Creek unites us; brings us together'.

Running somewhat parallel to the river, a 12 km stretch of Wattamolla Road twists and turns as it winds its way down from Berry Mountain towards the heart of Kangaroo Valley. The road connects the community; almost all the houses branch off from this main road. With steep slopes on either side of the road, it is relatively easy for neighbours to see weeds emerge and become established. There is also a sense of closeness brought about from the steep valley sides that turn into a rocky escarpment and make mobile phone reception patchy at best. Wattamolla Road has been described as, 'essential, beautiful, our "common ground" linking us, geographically and socially'.

The river and road that connect the community are key to understanding how and why a weed, which has little regard for boundaries, helped the community to bond. As with other examples of collective action, it begins with a shared problem and a willingness to address it together.

The Privet Poisoners Picnic identified that there were residents who were concerned about privet flowering along the riparian zones of Brogers Creek and its tributaries and who were keen to work together across private property, Crown and Council land. Over time, the group has grown such that there are now 20 to 30 people who regularly attend each working bee. They concentrate their efforts on managing weeds on private properties owned by group members; 'neighbours helping neighbours with weed control'. However, there are exceptions for landholders who are active in the community and may need a hand because of illness or other compassionate reasons.

This willingness to help others speaks to a broader aim that has formed part of the group's ethic almost since the beginning. In getting together to control privet, the small group:

> quickly realised that these afforded important opportunities to meet and chat and get to know each other. We gave ourselves a

> very wide brief to both protect our magnificent environment and nurture community.

The group builds community among members by holding barbecues and afternoon teas immediately after a working bee, providing an opportunity to socialise, to reflect on what has been achieved by working together, discuss what needs to be done next or any other pressing issues facing the community.

The importance of the post-working group food sharing is central to the success of the group. It draws people to Landcare, makes them more willing to participate in working bee activities, and provides them with a sense of connection to the group.

> Our post-Landcare reception and review will usually be longer than two hours! Sometimes it might [be], 'Now, we'll have a cake, have a cup of tea, have a beer and go home.' Sometimes they just roll on ... So, it is definitely a huge part of Landcare, having that post-get-together, because it's another reason that people might want to come along.

The discussions that take place are also important for keeping a finger on the pulse of the community and staying informed about current issues. Some describe the post-working bee discussion as a 'forum' that provides 'a good way to understand the issues that are facing landholders living in this part of the valley'.

In fact, through these post-working bee festivities and discussions, the group realised that there would be significant value in including the wider community. Now they host two events for all residents of the Wattamolla valley each year.

Since 1999,[11] the group has hosted an annual community picnic in January on the grounds of the old Wattamolla Schoolhouse. Everyone in the valley – connected by the river, road and weeds – is invited, and often there are 50 to 100 people who congregate, share locally grown produce and home-made cakes, listen to live music and enjoy games, such as the one-of-a-kind Australian Stone Skipping Championship. Kids love the festival atmosphere, with the giant tent, oversized cushions and whirligig swing.

To make sure that the picnic would be as inclusive as possible, Andrew approached the three long-term farming families to see if they would come to the inaugural picnic. The farming families all accepted the invitation, and the event has grown over time: 'In the picnic, we sit sort of where you are on Wattamolla Road, and everyone introduces themselves, and we have a sort of community forum.'

The forum provides an opportunity for residents to update one another on happenings in the community, to celebrate significant milestones and commiserate when residents pass. The topics over the years have included discussions about the state and safety of Wattamolla Road, the possibility of establishing a local renewable energy network, managing pest animals, protecting wombats, organising a Clean Up Australia Day event and assembling Team Wattamolla to compete in the Kangaroo Valley Show. Landcare activities are promoted each year and the invitation re-extended to all residents to join. As with other social events, the formalities are time limited, to make sure there is plenty of time for socialising.

In 2010, the community picnic provided a means for organising a working party who would work on a book about the communities of Brogers Creek, Wattamolla and Woodhill. Residents in the area were invited to contribute two pages each. In the final version of the book, there were 79 contributions, mostly first-hand accounts of people living in the community. The book was titled *Views and Visions: Stories of Wattamolla and Woodhill*. It was published by Brogers Creek Landcare in 2011 and, fittingly, was launched at the community picnic. The book is now in its third reprint and is used as way to invite new residents to the community, to join the Landcare group and the monthly working bees: 'Andrew was one of the first to pull up in his ute to give us the Wattamolla Views and Visions book, and say, "Come along to Landcare".'

Since 2017, COVID-19 permitting, the Landcare group has added a mid-winter feast to the social events that it runs for the community. The feast called Locavore is held at a different resident's home each year and involves inviting residents to come together and share produce that they have grown or made. It is another opportunity for residents to bond and get to know one another socially while also talking about issues of concern, such as preparing for the bushfire season.

The 2024 Locavore had over 70 people in attendance. There were fairy lights that drew people towards the five fire pits, and an abundance of food; spicy pumpkin soup, slow-cooked lamb and brisket, lemon tart, orange cumquat cake, to name just a few. The lemon poppyseed cake I baked (see recipe at the end of Part 1) went well with the citrus desserts. That night, the group celebrated their community spirit, a symbol of the resilience of this community to the many challenges it had faced in the preceding years.

The benefits of a belonging to a Landcare community

According to members, Brogers Creek Landcare has been successful in building community, educating residents about weeds and keeping environmental weeds in check along the river and the road.

The success of the group in nurturing community is evident in the accounts of residents who have recently moved to the valley and joined the group. When asked about their impressions of their first working bee, one responded:

> I thought it was great. It was very friendly, and I felt very welcome, and instantly I felt very at home with everybody ... And if not for the Landcare, I don't know how I would have met my neighbours, really.

Newer residents also value the knowledge that Landcare members willingly share about the area and how to manage land:

> I never knew what [wild] tobacco weed was, or privet, really, so the awareness. And the knowledge, again, is so valuable, that anybody will help you decipher anything that's a mystery.

Indeed, community education has been a core group activity for over 20 years; in 2001 the group purchased and distributed 100 copies of the booklet 'Weeds of the South Coast' to landholders in the valley.

They also produced a series of 12 'Weed of the Month' colour posters that were displayed at community noticeboards around Kangaroo Valley.

Beyond weeds education, the Landcare group aims to build ecological literacy. They have invited guest speakers to talk to the group about environmental topics.

> Every so often we have a guest speaker ... We had a really good riparian habitats talk one time. We turned up at Landcare and instead of working, we were just sitting by the river and Garry Daly was talking about the riparian habitats and how to conserve it ... That sort of thing is so useful, I think, having expert ecologists and things talking to us about various aspects of the environment.

The group has also used funding provided by Shoalhaven City Council, Sydney Catchment Authority (now part of WaterNSW) and NSW Environmental Trust to build knowledge about weed management practices. They have used external funds to lead trials to evaluate whether there are more effective ways to control dense privet infestations in their local area. The trials enable members to learn by doing.

With respect to controlling weeds, members have personally benefited from having working bees on their own properties. While this often only includes members of the Landcare group, there have also been working bees where students from the University of Wollongong Green Team, high school students and a council work-for-the-dole team have helped with weed control work.

One member showed me their visitors' book in which there appeared a written entry stating, 'First Brogers Creek Landcare at our new place. What a wonderful community we live in! Significant amounts of privet and wild tobacco were removed.'

Above the text is a photo of the team of volunteers who worked on the land that day. There are 17 people in the photo. Most wearing hats, some wearing gloves or holding pieces of equipment. In front of them is a pile of weeds, presumably their day's work.

For some, whose properties haven't had working bees, the benefits they see in removal of weeds applies to the broader area. They can see that the Wattamolla valley has fewer weeds than other places nearby:

> When I drive to other areas, I realise just how much is being achieved by our group, because sometimes we have working bees along the road and things, on the roadside, not just people's properties. When you drive to other places there's just weeds, chock-a-block everywhere, along the roadsides, I realise that I'm very lucky to be in this group.

Beyond weeds, the communication channels established via the Landcare group were highly valued by all members I interviewed. These included communications via email and WhatsApp. It might seem odd to mention something as mundane as this, however such communications were pivotal to surviving the 2019–2020 Black Summer bushfires and have enabled residents to manage multiple challenges that have affected the community since then, such as the COVID-19 pandemic and the unprecedented rainfall that three successive La Niña events brought. Wattamolla Road has been cut off at various times due to road slips and fallen trees. Given the number of other similar issues that have occurred across the region, the local council is not always quick to respond. And so, the group uses its neighbourly networks to notify residents and come up with local solutions where possible.

> That is another social benefit because if there's a tree that's come down, somebody will put it on WhatsApp and we all know that the road's closed, and then invariably somebody, in the group, will go and open up the road using a chainsaw.

Mad about Madeira vine

While privet brought the group together and provided an initial focus for working bees and community education, over time the group broadened its focus to include other environmental weeds, especially weeds that impact native vegetation. Madeira vine (*Anredera cordifolia*) is one such weed.

The way the group has worked together to control the vine, and the way it has eluded them, speaks to a common challenge among Landcare groups: what is achievable with regards to controlling weeds?

In 2002, Brogers Creek Landcare identified that there was Madeira vine (also known as lamb's tails) at three sites at the bottom of the Brogers Creek catchment, which they proactively removed through volunteer effort and the use of contractors. They subsequently identified a further four sites, one of which was on Crown Land. They worked together with Shoalhaven City Council to address this weed.

Like privet, the spread of Madeira vine is closely related to changes in land use. An article in the *Kangaroo Valley Voice*, the local newspaper in the broader Kangaroo Valley, in 2005 described Madeira vine as 'the worst environmental weed in the Shoalhaven' and was associated with removal of grazing from 'marginal lands, roadsides and creek banks'.[12] There was concern about how it was 'expanding rapidly and starting to cost large amounts of money and time to control'.

In 2005, seven sites on private property in lower Brogers Creek continued to have Madeira vine and were targeted through working bees. The Landcare group received funding from the Sydney Catchment Authority to pay contractors to help them control this weed.

Madeira vine is a particularly time-consuming weed to control. Its tubers need to be handpicked and carefully bagged. Any parts of the plant that are left behind can readily resprout. Members of the group described it as 'tedious; best to work in pairs with someone whom you want to chat with'.[13]

One of the group members reflects on the value of the Landcare group's efforts for raising awareness of this problem plant: 'I remember you talking to us first about Madeira vine, and we'd never heard of it.'

A map of the status of efforts to control Madeira vine around Kangaroo Valley in December 2008 reveals that the efforts of Brogers Creek Landcare dovetail with those of the Kangaroo Valley Environment Group, which spans the whole valley, including Upper Kangaroo Valley Landcare (Chapter 4). The map identifies 51 Madeira vine sites along the valley, of which only four were considered eliminated, two of which were on Wattamolla Road and another nearby. Three of the remaining four sites along Wattamolla Road were 'largely under control'.

In 2012, Madeira vine was listed as a Weed of National Significance. It is 'one of the worst weeds in Australia because of its invasiveness, potential for spread, and economic and environmental impacts'.[14]

Yet the early efforts of the Landcare group, which pre-dated the listing of Madeira vine as a Weed of National Significance in 2012, have most likely only slowed its spread:

> It's not doing nothing, but on the other hand it isn't doing everything that needs to be done because we can't. As a small group of people, the extent of the problem is beyond us. But we are making some impact.

In the early 2020s, Madeira vine was still considered a weed of concern for Brogers Creek Landcare. Although they've only got small patches of it growing, some members consider it to be the worst weed they have.

It is significant that Brogers Creek Landcare tackles Madeira vine and multiple other weeds. Unlike some groups like Tapitallee Budgong Local Landcare (Chapter 5) and Tathra Land Care (Chapter 7), which predominantly focus on single weed species, Brogers Creek Landcare focuses on a range of environmental weeds. This is consistent with other groups that focus on protecting riparian zones, such as Upper Kangaroo Valley Landcare, Deua Rivercare (Chapter 6) and Candelo and District Landcare (Chapter 8). The desire to maintain and encourage native vegetation along a river corridor, often requires a coordinated multi-weed strategy. Such a strategy means that the group is likely to take longer to see progress in controlling individual weed species. However, their efforts may help restore specific ecological communities.

Making progress on multiple weeds?

It had been an hour of hard yaga (often written as 'yakka', yaga is a Yagara word from the Brisbane, Queensland area that means work).[15] I had been working my way up the steep slope that rises sharply from

Wattamolla Road. I had started by focusing on the wild tobacco (*Solanum mauritianum*), which is easy to spot with its furry grey-green branches and leaves that tower over the surrounding vegetation. I was covered in the sticky seeds of farmer's friend (*Bidens pilosa*) that had seen fit to stick to my jumper, my jeans and most of my hair as I had pushed through shrubs to cut down the wild tobacco. I had initially felt satisfaction at cutting and painting and watching the large leaves of the wild tobacco plants wilt as their trunks slowly reacted to being cut from the base.

It was only as I continued my way up the hill that I noticed the moth vine (*Araujia sericifera*) pods. You would think with their size that you could easily spot them from afar, but somehow, they are easily disguised and usually there is only one pod peeking through shrubs that gives them away. And once you've found one, you know there will be more.

As I looked satisfied upon my pile of pods someone asked me whether I had seen the cassia (*Senna pendula* var. *glabrata*). I had totally missed them in my focus on the wild tobacco and moth vine. I looked back down the hill and spotted their bright yellow flowers. How had I missed those? I had only just learned about them this morning at a Deua Rivercare working bee. And of course, once I had spotted one, I began seeing them everywhere.

Another Landcare member nearby was slowly making her way through the lantana. The task at hand seemed insurmountable. The weeds went on and on, up and up the hillside – 12 of us labouring was barely going to make a dent on someone else's property. Someone who apparently didn't do much to control weeds.

I began to appreciate why many get disheartened with weeding. Why some people don't even begin and don't join groups like these, because if they did when would the weeding end? It also made me appreciate why some groups focus on a single weed, it makes it easier to cover a larger area and feel a sense of progress.

I was feeling overwhelmed. What was the point of continuing? What difference was I really making?

I stopped.

I made my way back to the cars, emptied my shoes of leeches, tried to shake the seeds from my hair and looked back at the hillside. The others were still chipping away at the weeds.

I started a conversation with the Landcare members. I wanted to understand why they persevered.

They explained that this is a highly visible hillside on Wattamolla Road. Many of them see the weeds as they travel along the road and know that if they don't help out, they will continue to be infested by these weeds.

While they work, they look for native seedlings. These emerging native trees act as beacons – something that they have a responsibility to help free from the weeds. They pointed out some native giant stinging trees (*Dendrocnide excelsa*) that they were particularly excited to see coming up through the weeds.

They also know the local landholders personally and the many reasons why they can't control the weeds. So for them, this is a social project as much as a weedy one. The Landcare members also know that if and when they need a hand with their weeds, the others will be happy to pitch in and help. Of course, there is the added benefit that working bees usually end with a cup of coffee and a cake.

I looked across the road and spotted some more moth vine pods. I knew that if I left them there it wouldn't be long before they ripened and released hundreds of seeds covered in silky white hairs that would be carried away by the wind. There was a task to be done that I could do right now, that would help the others. It mightn't make much of a difference in the grand scheme of things, but it was better than doing nothing. Feeling reinvigorated, I put my gloves back on and got back to work.

As I drove home that evening, still pulling weed seeds out of my hair and off my clothes, I reflected on what had motivated me to continue weeding after I had become disheartened. I knew that if I had been alone, I would have walked away from it. But there was something about watching the others continue to work, a sense of respect and responsibility to them, that inspired me. Even though I had only met them a few times, I had come to learn how important weeding is for the group, for bringing people together and making a difference to the local environment. When I stopped weeding and talked to the others, I had a chance to learn about how weeding fits into a bigger vision for the valley that the group had been working towards since the 1990s. The group is about much more than weeding or caring for the local environment. Making sure people feel included and building strong friendships is clearly part of the 'secret' to their success.

Taking the stigma out of weeds

Beyond motivating people to control weeds, Brogers Creek Landcare has overcome social norms that discourage rural residents from talking to one another about weeds.

As mentioned in Chapter 1, rural social norms deter neighbours from talking about weeds so as not to cause offence. Other rural social norms discourage neighbours from making comments that could be considered as interfering in the way others manage their land.[16] These norms are reinforced by expectations that good landholders are those who know about and control weeds on their own.

Such social norms and expectations make it surprising that Brogers Creek Landcare has created a space where members can comment on one another's weed problems, ask for help and help one another out.

Over the years, group members have found a way to make themselves vulnerable to one another and 'giv[e] each other permission to have a critique of our land practices'. Along the length of the Wattamolla Road, 'people put their hand up. "Yep. I need help. Come and help me next time."' In doing so, they reap the benefits of not having to weed alone.

> For the three years we've been here, I think we've had three times where the Brogers Creek Landcare group have come in and it's fantastic, but we get huge swathes of area of privet removed, and it raises our hearts just to see a significant amount of area cleared in the two hours that they're here to actually do stuff.

The Brogers Creek Landcare group has normalised having weeds. In this valley everyone recognises that they need help. 'None of us have finished our weeds. We're not coming along because we've got no weeds to do!' 'Look in any direction, and you'll find a [weed] problem.'

The group makes it acceptable to own up to having weeds because of the recognition that weeds are a shared problem. The group also makes talking about weeds palatable by focusing on the benefits that come from controlling weeds together. An article in the *Kangaroo Valley Voice* reveals the way that the group motivates itself to keep going, by focusing on what they have achieved collectively.

> On this day, there were a few grumbles - a tractor could do this job in no time. What about these shrubs - do they go too? People were a bit hesitant at the start - how do we do this? But a few souls started to cut into it [the dead lantana], everyone got enthused and bingo ... What a treat, transforming the landscape, one paddock at a time.[17]

So the way in which Brogers Creek Landcare has built capacity to control weeds over time has been by forming strong social relationships that make it possible destigmatise having weeds. They also focus on the bigger picture: on the environmental benefits that come from controlling weeds and on how much more they can achieve together than individually. These are not one-off activities, it requires long-term commitment to the environment and one another, through regular and repeated actions, month after month, year after year.

Don't tip-toe around tensions

Research on collective action suggests that managing conflict is a big challenge associated with getting people to work together.[18,19] Here again, the Brogers Creek Landcare group provides a rich example of how a group negotiates the inevitable tensions that come from controlling weeds in a rural community.

One of the key tensions that exists for the group is around the use of herbicides to control weeds, what is known as 'the glyphosate tension'. The group are caught in a middle ground between those in the community who have a zero tolerance of herbicides and those who believe that it should be applied more widely to control weeds quickly and effectively. So how do they answer the question: 'to use or not to use glyphosate?'

Some members of the broader community are uncomfortable with using herbicides and believe that the group is too liberal in its application of glyphosate, believing that '[they] get the boom spray in on the road and just spray a bunch of glyphosate in the gully to kill all the lantana'. Such concerns reflect growing concerns globally about glyphosate[20] and they have influenced the decision of those residents not to become involved in the group:

> There was some conflict a few years ago ... some people who had lived there for a really long time were really not happy about having chemical herbicide applied to control lantana on their property.

These residents felt 'ostracised' and as though they couldn't 'meaningfully participate in the Brogers Creek group'. The reasons why such landholders don't want to use glyphosate can be seen in the comments made by a landholder in a nearby area:

> Because of the cloud that's hanging over the use of RoundUp® in particular, it would be nice to know that, if it's safe, because each article you read has got a different perspective on whether it's safe or not, and also the concentrations.

Such concerns are recognised among group members: 'some people don't come, or come less, because they don't want to be in a glyphosate [group]'. Yet the group's preference is to use glyphosate, because they believe it is the most effective way of controlling weeds like privet:

> We had one working group for the [neighbours] here, where we didn't use glyph[osate], but we cut huge amounts of privet and burnt it, but it was soul-destroying, because in the next few months, no matter how low you cut it, you cut one and you ended up with twenty.

This is a recurrent theme across the Landcare groups. Often new members and rural residents are reluctant to use herbicides when they first start weeding but over time come to accept that judicious use of herbicides can make an appreciable difference to weed control efforts.

In the case of Brogers Creek Landcare, their commitment to building community and being inclusive means that they have tried to accommodate diverse perspectives on herbicides by not requiring group members to use herbicides, 'with glyphosate, not everyone has to use it, because in the working bees you can do whatever you feel comfortable with'.

Because of the group's considered and deliberate acceptance of anyone willing to help with their collective weed control effort, some landholders in the community have joined the group on the understanding that they will control weeds by pulling and cutting.

> I don't actually use poisons, so when we go along, the others are using glyphosate, but I just cut and pull ... I'm not pressured to use them when I go to Landcare. I think they're happy for me not to.

At the other end of the spectrum, some in the farming community believe that the Landcare group doesn't use herbicides, 'I think they don't like to use chemicals as much'. It seems that this perception may have arisen because the Landcare group includes 'lifestylers' who don't 'want to use any sort of chemicals'. It also comes from a perception that cutting and painting with herbicide is highly labour-intensive and too time-consuming. So those who prefer quicker forms of weed control that cover a bigger area, like spraying, are also unwilling to join the Landcare group.

One of the other key tensions in the group relates to where they focus their weed control efforts. In part, this is a question about whether to focus only on land that has environmental significance or also to include agricultural land. For those who prioritise the riparian zone and natural bushland, they may opt out of working bees that involve picking fireweed out of paddocks. However, the bigger and more challenging question is about where to hold working bees; whether they should only be held on the properties of members or whether they should also be used to help others in the broader community who aren't active in the group.

> They used to argue against working on places where people weren't effective, you know, and a couple of other people like that. I was quite happy to do it; I saw it as strategic, but other people in the group would say, 'No, we're not going to work on them, because they're not coming along and helping.' So there always was that tension.

One of the reasons that this tension arises is because the idea of reciprocity is central to the ethic of the group. The choice of where to hold working bees is as much about helping those who have helped others as it is about making oneself vulnerable to the rest of the group by admitting that you need help with your weeds and then benefiting from having the help of others.

And so the group often makes difficult decisions about where to hold their working bees. In the research on collective action, clearly defined boundaries are seen to be one of the key ingredients to getting people to work together on joint environmental problems.[21,22] Here we see that not only does the group limit its activities to the Wattamolla valley and the 12 km stretch of road that connects the communities but that it mostly limits its activities to the properties of group members and the adjoining roadsides.

> Member 1: But there are limits on what we're prepared to do.
> Member 2: We won't go to someone's place if they don't join. Occasionally, people will say, 'Could your Landcare group come and get rid of weeds for us?' But they've never actually attended a working bee, so that's one. You have to be part of it.

And yet there are exceptions. There are some landholders who need a hand with their weeds who are not members of the Landcare group but are deemed worthy of a working bee because they contribute to the community in other ways, such as volunteering for other key social groups, or because they are elderly or infirm. In such situations, the community-oriented ethic is used to negotiate this tension.

> And next weekend we're going to be at the [non-member's property]. They're old, and their son just died two weeks ago. And it's sort of an act of ... kindness for the family. And we've done that for the [non-member's property]. Just as a way of the community saying, 'We appreciate you' and give people a lift.

The internal logic of how working bees are selected is not clear to those outside the group. For those who believe the main aim of the group is to reduce weeds, they believe that the group is not being sufficiently strategic.

> When they do have a working bee, they will get a lot of work done in a very small area. I wouldn't say that when they have a working bee they look at a collection of property ideas. For instance, I'm not seeing Brogers Creek [Landcare], for instance, targeting privet in that linear concept across properties ... it's not a holistic approach.

For outsiders, it is not clear that the group makes decisions based on its dual aims of controlling weeds across the landscape while also nurturing community. While this may discourage some people from joining the group, for those who are members such decisions are critical to their ongoing membership.

> There's been times when there's been elderly couples that have been losing control of their weeds and things, and we've done a session there, which I think is quite a good community thing to support people that are having a hard time with it.

The way in which the group consciously negotiates the tensions that it faces is one of the things that binds it together and may explain why it is one of the groups in this book whose membership numbers are increasing. Indeed, those familiar with the group believe that the need to resolve tensions helps Brogers Creek Landcare thrive.

> The Brogers Creek group, it ticks over because of the tension. That might be a little bit of a funny thing to say, but if everything was kind of happy-go-lucky and merry all the time, you can kind of get a little bit lost in that. I think you need to be kept on the edge and a little bit on your toes ... A commitment from participants to listening and sharing information, especially about their fears and anxieties over how and why weeds are managed in a particular place at a particular time. I think that's a really healthy thing, and if that happens and it just brings enough edge to the conversation to keep things ticking over, then I think you'd have a really sustainable and really dynamic group.

Ultimately, the group recognises that they will never achieve 100 per cent membership among the residents of the valley. They know that there are people who move to the valley 'to get away from people' and who don't want to use glyphosate, as well as the original farmers who are 'busy doing their real farming' and those who pride themselves on being able to 'look after my own place'. They also are pragmatic about what is achievable when it comes to weeds. For this reason, their dual aims are important. A two-hour working bee on the last Saturday of every month allows them to achieve both: progress on weeds and a space to connect.

What next?

Brogers Creek Landcare is undoubtedly thriving. The numbers it regularly gets at working bees are a source of inspiration for other groups, not only with regards to the absolute number of people who come along but also the spread of ages. I have seen kids as young as 15 chopping down wild tobacco, and university-aged adults, who used to come along as children, now bringing their friends to attack weeds when they're home on Landcare weekends. There is also a mix of longer-term and newer residents who come along. Those who live in Sydney make sure they are in Wattamolla for working bees.

I know the leadership team is continuously looking for new activities to participate in and opportunities to grow the group's network with other organisations. And for now, there seems to be enthusiasm among the group to keep trying new things and finding new activities to keep the group connected.

Given the vibrancy of the gatherings, this group gives me hope that they've found the right formula for this valley and could still be here for another 30 years.

4

Bushcare and Landcare in Upper Kangaroo River

There are two things you can rely on if you join Upper Kangaroo Valley Landcare for a working bee. One, that there will be plenty of laughter as you make your way through the beautiful bush that surrounds the Upper Kangaroo River. So much so that you may need to wipe a tear from your eye so that you can spot one of the ever-diminishing weeds that draw the group together. Two, that there will be an abundance of cakes, biscuits and other home-made delights when morning tea rolls around.

Upper Kangaroo Valley Landcare is not just a group of volunteers who come together once a month to pull riparian weeds. They are a group of friends who help and support one another in this relatively isolated community. They offer to pick kids up and drop them off at the school bus stop, jointly volunteer for a range of arts, environmental and community groups, put together a stall at the annual Kangaroo Valley Show and host the occasional weed walk or movie night for the broader community. And they've been doing this for decades.

The heart of Upper Kangaroo River: a hall, a footbridge and a river

The Upper Kangaroo River Community Hall provides a focal point for the small community. It is a quaint grey weatherboard building that proudly states its name and the year it was built: 1963. To this day, it is used to host movie nights, musicians, pop-up artist workshops and occasionally weddings, making it a place that is important for the residents of Upper Kangaroo River and beyond.

The hall is across the road from a 50-m-long suspension footbridge: a structure that has been built and rebuilt since 1907 to enable residents to cross the river and for children to get to the former Upper Kangaroo River school.[1]

The footbridge is also a drawcard for tourists. It has featured in tourism advertising since the mid-1920s,[2] when it was described as a 'quaint suspension bridge'.[3] In 1981, an article in the *Australian Women's Weekly* described the footbridge as 'a photographer's delight'.[4] Even today, it is not unusual to see cars and campervans parked next to the bridge on a grassy verge ideal for picnics.

For locals, the footbridge is special because it provides a great vantage point for spotting platypus, especially early in the morning, although I'm yet to be fortunate enough to see one.

The river alongside the footbridge is also significant for the Landcare group. It is the place where Upper Kangaroo Valley Landcare came into existence. Legend has it that one fine day in 1995 a dozen residents met at the Upper River Hall. The main agenda item for the day was finding volunteers for maintenance of the then 32-year-old hall. While discussing what needed to be done, one farming resident saw something across the road that concerned her more than the hall: small-leaf privet (*Ligustrum sinense*) in full bloom. As someone with a horticulture certificate, she knew that to protect the 'natural rural river, its vegetation and wildlife'[5] they needed to nip the privet in the bud, literally. That very same day she recruited a handful of volunteers and they got straight to work. The hall maintenance would have to wait.

From Bushcare to Landcare

From that first informal working bee, the residents decided to form a Bushcare group so that they could formally work on council-owned land.[6] A member of staff from Shoalhaven City Council visited Upper Kangaroo River and gave the residents some sage advice about weeding on public land. He advised the group to limit their work to an area that would be achievable, taking into account the size of the group and the extent of the weed challenge. The group took the advice to heart and focused their weed control efforts on the riverbanks across from the Upper River Hall no further than Gerringong Creek Road, 200 m downstream from the hall.

This is a place the group has continued to come back to over 30 years, although less often now than at the beginning. Not only have members spent countless hours removing weeds but they have also replanted it with native vegetation, including red cedars, which were frequently logged throughout the early 1800s.[7,8]

Historic photos of the footbridge reveal an almost unrecognisable riparian zone. There were few trees to be seen on the riverbank opposite the hall, and the riverbank closest to the hall had only a smattering of eucalypts. The cleared paddocks, felled trunks and wooden fences speak to widespread logging in the valley.

Today there is dense wooded canopy that hides the bridge from view. Thanks to the Landcare group, the riparian vegetation is largely free of weeds. The few privet left have little to no chance of becoming established. Residents appreciate that the banks are weed-free thanks to the group. One member reflected:

> Over the years I've seen that group make dramatic changes in areas, particularly around the hall, around the river around the hall area, that was transformed by the earlier members of the Landcare group.

During the first few years as a Bushcare group, the volunteers came together for two hours once a month to remove privet from the river. Over time, they couldn't help but notice that they were treating the symptoms rather than the cause. They could see that privet was seeding upstream of the hall, and that if they were ever to make a long-term difference to the river, they would need to start working on private properties too. This would require them to become a Landcare group.

It wasn't an easy decision to make. There were concerns about the administrative burden required to start a Landcare group. To be able to apply for grants on private property the Landcare group would have to be registered as an incorporated body. With incorporation comes annual meetings, reports, insurance and the need for volunteers to commit to leadership roles. Others in the community were keen on the idea of a Landcare group but were hesitant to commit to additional meetings. The sentiment is still true today:

> Nobody will take over being a treasurer! God, I don't blame them. There's got to be somebody. There's got to be someone who's kind of okay with numbers, or is happy to do admin work. I'd rather be pulling out the weeds, thank you.

Despite the hurdles, in February 2003 the group took the first steps towards forming an incorporated Landcare group. Not only did they begin to work out the administrative details but they also initiated discussions about what area they would work on and what weeds to include. They wanted to make sure that what they were committing to what was achievable. The then NSW Department of Infrastructure, Planning and Natural Resources helped them identify a list of priority weeds and effective management strategies.

The group also recognised that they would need start-up funds to purchase insurance and tools to do the work. To raise these initial funds, they decided to invite all landholders in the area to volunteer or become financial members of the group. This doubled the reach of the group, with equal numbers of active and financial members. Not only did this enable the group to raise the funds needed for the initial expenses but the broad membership base was crucial to obtaining grant funding down the track.

Upper Kangaroo Valley Landcare formally commenced in October 2004, almost 10 years on from the first efforts to control privet across from the community hall. It began with two related projects on the Upper Kangaroo River: controlling privet upstream of Gerringong Creek and Madeira vine (*Anredera cordifolia*) downstream of Flat Rock. These projects delimitated a 5 km stretch of river, which the group has continued to keep as its focus area for all activities since then.

One of the things that is curious about the minutes of the first official meeting is that no-one was appointed Chairperson. It was unanimously agreed that the role of chairperson would be rotated among members. The residents who were voted secretary and treasurer would continue to hold those roles for many years to come.

What is also interesting about the way Upper Kangaroo Valley Landcare formed is that the Bushcare group didn't immediately dissolve. For many years afterwards two working bees were held: one for the Landcare group on the third Saturday of the month and another for the Bushcare group on the first Friday of the month.[9] Being part of both Bushcare and Landcare has provided the group with additional support from the local council as well as flexibility in terms of working on public and private land.

Working bees today

In May 2022 I visited Upper Kangaroo Valley Landcare for their regular working bee and Annual General Meeting. It took four attempts to join the group thanks to COVID-19 precautions and 'The Slip' – a landslip that cut off access to the upper part of Upper Kangaroo River Road.

That day, I made the long drive into Kangaroo Valley via Berry, over the historic Hampden Bridge, and turned right onto Upper Kangaroo River Road. Past grassy fields, over cowpats that draw attention to the working farms still operating in this part of the valley, under a kookaburra sitting on the telephone wires and through a section where vegetation encloses the road. Depending on the time of year, the purple flowers of wild tobacco (*Solanum mauritianum*) add a splash of weedy colour to the lush green landscape.

And then I reached The Slip. There were traffic controls in place but it was driveable if you went very slowly over the sizeable ruts and potholes.

From The Slip there is a good vantage point over the Upper Kangaroo River. You can see the river sheoaks (*Casuarina cunninghamiana*) that have their feet in the water, holding tight to the sandy banks. On a sunny day, the rocks invite you to come down and dip your feet in the almost clear water.

Three kilometres further up the road is the Upper River Hall.

I was the first one there that morning, just before 9 a.m. I barely had a moment to gather my thoughts before other cars started arriving and parking nearby. As people emerged from their cars, there was a buzz about them. Cakes were produced and taken to the hall in preparation for the Annual General Meeting that was scheduled after the working bee. The mouth-watering treats provided strong inspiration for the weedy work to be done.

The usual protocol for the working bee that precedes the Annual General Meeting is to work for an hour or so along the banks of the river, directly across from the hall, just up and downstream from the footbridge. This working bee is as much a nod to the beginnings of the group as it is a demonstration of ongoing dedication to looking after this locally important public space.

At each working bee, the back of a ute is filled with tools and equipment: battery-powered chainsaws, bags for collecting weedy matter, tool belts that carry secateurs and small bottles of herbicide. Later, the ute will be filled with bags of weeds, ready for a trip to the tip or someone's burn pile.

I fetched my gloves, introduced myself to the volunteers and grabbed what I needed from the back of the ute. One of the volunteers relished telling everyone about the leeches and ticks that she had encountered recently and so we all diligently applied insect repellent, keen to avoid such encounters.

Before we had strayed far, two of the volunteers spotted moth vine (*Araujia sericifera*) at the edge of the roadside and therein began the weedy work.

We split into small groups of three or four to work along the river, some along the banks, others further up, never much more than a few metres apart.

In our group there were plants we were confident in identifying, like moth vine, but other spiky plants we weren't so sure about. To be certain, we would take a leaf with us, and make a mental note of where the unidentified plant was, before tracking down one of the recognised experts in the group for advice about whether the plant was a 'goodie' (a native plant) or a 'baddie' (a weed of concern). And that's one reason why many people join Landcare groups in Kangaroo Valley and elsewhere, because there are local experts on hand who can help tell a weed from a look-a-like.

In the meantime, the other group had found a Madeira vine. Four people slowly and delicately worked to disentangle it from the other vegetation it had grown over. I had never seen weedy work done so carefully and meticulously before. I learnt that day that Madeira vine is a plant that resprouts very easily from any part left behind, be it leaves, tubers or roots. So the volunteers were tenderly removing the plant and carefully placing each part into a bag.

High above us we could see moth vines that had grown out of reach, 20 to 30 pods taunting us from on high. You would need a tall ladder or a cherry picker to reach them. We stood on the road to discuss the equipment or contractors needed to remove the moth vine pods, deciding it was time to down tools for the day.

We all made our way back to the hall for the Annual General Meeting. More food emerged, along with tea and coffee, until the table was overflowing with home-made treats. Thankfully some more people turned up for the meeting to help make a dent on the plates of arancini balls, fruit, cakes and biscuits.

The meeting celebrated the number of hours that the members had spent volunteering over the last year – 237 hours in all – supported by a further 165 hours of contractor time. There were updates on the projects undertaken. This year the projects had included Bushfire Recovery funding to manage weeds and control feral animals as well as a project targeting Chilean jasmine (*Mandevilla laxa*), which the group had first become concerned about in the early 2010s.

There was discussion about when working bees are held and the potential impact this has on membership numbers. Does holding them on a Monday exclude weekenders from coming? Is it worth trying Saturdays? The group was split as to whether it was a good idea or not. Fridays were ruled out, having previously experienced poor attendance because residents are often busy preparing for visitors.

While a few of the groups discussed in this book have wondered about whether they should change days, Upper Kangaroo Valley Landcare is the only one that has done so to see its impact on volunteer numbers. During 2022 and 2023 the group experimented with Saturday working bees. Unfortunately, attendance got worse, rather than better, meaning Mondays were reinstated as the best day to meet.

At the meeting, the group also discussed where the next working bees would be held. One of the members offered to host it at their place, which is a '[wild] tobacco forest', but the idea was turned down because the property was located downstream of Gerringong Creek Road. Given this group started by working on council land then expanding its focus to private properties, the question of where to focus efforts was more actively debated than in the other groups discussed in this book; it was discussed at all four Annual General Meetings I attended.

Clearly defined boundaries

Research on collective action has repeatedly found that the first thing cooperative groups need are clearly defined boundaries of what area the group will cover and whose cooperation is needed.[10,11] While defining boundaries sounds straightforward, for problems like weeds it is hard to say where the problem starts and ends.[12] The case of the Upper Kangaroo Valley Landcare shows some of the reasons why it can be challenging. Here, as elsewhere, the group has long deliberated how to find a balance between defining an area small enough that it is feasible to make a difference to the weed problem, but large enough that it can recruit sufficient volunteers.

Having clear geographical features, such as a valley, or a change in land use like a national park, can be useful for limiting the reach of a group. Such features are key to the boundaries that define where Brogers Creek Landcare, bounded by the Wattamolla valley and connected by Wattamolla Road, and Tathra Land Care, bounded by two national parks and an ecological reserve, work. In this regard, the choice of stopping their work at Gerringong Creek Road seems logical for Upper Kangaroo Valley Landcare. It is just before The Slip, after which point the valley opens into grazing land. As one interviewee explained:

> They [the Landcare members] always just went, 'We're riparian, and we're not going beyond Gerringong Creek.' They've stuck to that boundary. And they stuck to the boundary because they foresaw there wouldn't be enough people to go further down the river, and thought, if we don't spread our resources too thinly, we can really concentrate on it ... the bridge being a good physical boundary that everyone can see, it's like that's where we stop.

Having worked on this area for almost 20 years, most of the members of the group agree that it meets the definition of being achievable.

> I found that was really important with the group, just to set our targets to what was achievable ... For them to feel like they are doing something worthwhile. They've got to be able to ... see the effect of what they're doing.

Others in the group agree that with only one two-hour working bee per month, you really need to keep the area tightly bounded. Not only is this important for keeping members motivated but it also determines whether others are willing to join: 'a lot of [residents] just don't take it seriously. And I think part of that is because they don't see it as being an achievable project.'

Like Tathra Land Care (Chapter 7), Upper Kangaroo Valley Landcare allocate smaller work areas within the clearly defined 5 km stretch of river along which the group works. It is important for members to know that when they see weeds within the designated boundaries, that at some point in the year the group will control the weeds, that they will be taken care of: 'You never think of the whole lot. It's far too daunting.'

To this end, there are eight or so regular spots where Upper Kangaroo Valley Landcare meet each year. This includes Billai Creek, Spring Willow, Parrishs Road, Scotts Road, Treefern Road, the Rainforest River Retreat, Flat Rock and the picnic grounds. During each working bee the group works along the road from the nominated spot for that month towards the next spot and back again.

One of the challenges in choosing the area upstream of Gerringong Creek Road as where the group will work is that it excludes properties of people who live downstream but feel part of the Upper Kangaroo River community and who are willing to give up their time to control weeds on public and private property.

For these residents, part of the frustration comes from how highly visible weeds are along the road downstream of The Slip. Wild tobacco lines much of the road and spreads into neighbouring properties. It is a regular reminder of the extent of the weed problem and amount of work needed. However, it also provides an important point of contrast with the more weed-free area upstream of The Slip. Downstream residents enjoy bringing visitors to the footbridge, free from the thickets of wild tobacco closer to home.

Mostly, downstream residents understand why the group doesn't expand its activities.

> I would love Landcare to expand down the Upper River Road as far as us, of course, but I think with the resources that we have, we stay in a pretty small area up the top of Upper River.

What is implicit in the discussion of boundaries are the questions of what weeds to work on and how many people are willing to contribute to weed control. Tathra Land Care are the most similar to the Upper Kangaroo Valley Landcare in terms of the size of the area on which they focus. For Tathra Land Care, they have found it is easier to control one weed than many. While Upper Kangaroo Valley Landcare was initially inspired to act because of privet, their evolving concern for Madeira vine, moth vine and wild tobacco means that there is wisdom in not expanding the work area given the size of their group.

Why work elsewhere?

No matter what you spend your time doing, there are always trade-offs: something else you could be doing, somewhere else you could be instead of weeding. This is a theme that is mentioned across many groups. In

Tathra, the sound of the waves might tempt you to go for a surf rather than join a working bee pulling out bitou bush (*Chrysanthemoides monilifera* ssp. *rotundata*). In Kangaroo Valley, it's not just the many different recreational opportunities – kayaking, horse riding, bushwalking and bike riding – that compete for one's time but also the weed work right outside your own back door or in the back paddock or in the communal land you share with your neighbours.

Multiple group members reflected on the trade-off they make when they decide to spend time volunteering for Landcare, rather than working on weeds on their own properties.

> Sometimes there's a bit of reluctance there, especially when people are working on someone else's property, and I think as it was even mentioned at the meeting, the comments, you know: 'I've got enough weeds at my place. I could be doing this at home!'

In part, the group's focus on riverine and roadside areas means that working bees are rarely held on member's properties. This is not only true for those who are outside of the focus area but also those who are within it: 'we don't work on private property, really. Sometimes if we get permission in certain places, we might, but we don't work on private property.' This distinguishes Upper Kangaroo Valley Landcare from other groups, like Deua (Chapter 6), Brogers Creek (Chapter 3) and Tapitallee Budgong (Chapter 5), where the focus is on private property and where members directly benefit from working bees.

The other challenge that this group faces with regard to the time trade-off is that some members live on multiple occupancy blocks, where they already have a responsibility to work with others on weeds across properties. Three residents of multiple occupancy blocks explained that they have existing monthly or quarterly working bees with their neighbours to manage weeds and other maintenance required on communal property. Thus, the work of Upper Kangaroo Valley Landcare competes with the demands of other groups of landholders. Far from being surrounded by residents who don't care about the weeds, the Landcare group is just one of many cooperative weed efforts in the valley.

Residents who don't work on multiple occupancy blocks indicated that they liked joining Landcare working bees because it gives them a sense of purpose and companionship.

> I'm much more motivated when I've got somebody else with me. So, for me to go and do two hours [of weeding] by myself would be really hard, but I only need one other person. I only need Belinda to say, 'Oh, can you help me with such-and-such?' and I'm sort of right in there ... It was one of my favourite things to do, for years. And I still enjoy it.

The importance of weeding together has been embraced in the latest efforts to increase Landcare volunteer hours. A new informal buddy system known as the Friends of Upper River Environment is being trialled, where members can make an occasional request via WhatsApp for others to join them when they have time available to weed along the roadside and riparian zone. In principle, this approach should make the most of enthusiasm for weeding and a desire to work together. And indeed, it has meant some much-needed weed work has continued even when there haven't been enough people for working bees to proceed.

To herbicide or not to herbicide

The use of herbicides is a hot topic in a community that is as environmentally minded as Kangaroo Valley. They have long been at the forefront of environmental programmes. In 2003, there was a push to make Kangaroo Valley the first plastic bag-free town in Australia. In 2008, tourist operators, along with many householders, farmers and schools pledged to make Kangaroo Valley carbon neutral.[13]

As with other Landcare groups in and around Kangaroo Valley, some in the Upper River community choose not to join the group because they do not support the use of herbicides.

> I came [to a Landcare working bee] and then was just mortified that what they were all doing – they all had cans of Roundup® to work with and I said, 'No, I can't do this, sorry. It's not what I was intending … I think I'll just go and work on my own property.'

Others are also averse to using herbicides but join the group once they understand how sparingly these are applied and why they are used.

This is an issue that the group has grappled with since its inception. A report on group activities from 2005 documents the reluctance of some to use herbicides and the rationale for doing so:

> Some spot spraying of the regenerating vines with herbicides where the tubers are inaccessible between rocks has been approved by Sydney Catchment Authority and the NSW EPA [Environmental Protection Authority] but the reluctance of some landholders for using chemicals means the manual task is very onerous or impossible. We fear some sites may not get the required maintenance which will cause further spread of regenerating tubers downstream and negate the work already performed.

More recently, group members have sought to alleviate concerns about herbicide by measuring the amount used by the group over the course of a year.

> Three years ago, I strictly measured how much we used for the year. Average, for, say, eight people per session, for twelve months, we used a cup. We used one cup of poison.

While the group uses very little herbicide during working bees, they recognise that herbicides make weed control easier for contractors and for controlling weeds on private property. In 2022, the group purchased, licensed and insured a 1,000 L spray unit to allow residents of the community the opportunity to spray weeds with herbicides on their own properties. With the investment in this trailer, it is clear the Landcare group sees herbicides as an integral part of managing weeds in Upper Kangaroo River. They know how much time and effort it takes to control weeds and see this as one of a range of approaches needed to make a difference.

Community education

According to an A-frame sign, Upper Kangaroo Valley Landcare does more than control weeds. It also works to control feral animals, slow stream-bank erosion, propagate and plant local native plants, and educate the community about the diverse native ecosystem. Of these broader aims, the one members are most proud of and dedicated to is community education. They are passionate about learning and sharing their knowledge about the local environment with others.

For many members, Landcare provides an opportunity to not only learn about weeds, but also native plants and animals.

> There were these two people who were just like walking encyclopaedias, so it was a way for me to learn about the land. It was very much an educational thing, for me. And very rewarding ... I think probably for most people, but certainly for me, that learning and the knowledge, having somebody knowledgeable to work with is really, really critical.

For another member, the opportunity to learn about the land through Landcare was tied into an ethics of care.

> I think a lot of us go there because it's a way of keeping fit and active as well, and also getting to know our patch of the world, too, and a sense of looking after – custodianship. Looking after that patch. Care. An ethics of care.

To this end, the group not only aims to educate residents about the environment but also the broader community. Since at least 2012, members of the group have volunteered their time at the annual Kangaroo Valley Show, helping at a stall run by the Kangaroo Valley Environment Group, which seeks to advertise and communicate environmental activities taking place in the valley. Other events the group has contributed include the 2014 Tails, Scales and Whales festival and a display and talk at the 2016 Quollidor Project, aiming to raise awareness about the spotted-tailed quoll (*Dasyurus maculatus*), a threatened carnivorous marsupial species in the area.[14]

Within Upper River, the group has hosted and contributed to a range of environmental events. In 2022, they held a 'Weed Walk' where several local experts – Aboriginal Elders, ecologists and consultants – were brought in to lead residents on a walk through the bush to identify plants and animals. Over 25 residents attended the walk, making it the second largest Landcare event of the year.

> I think you can sleep easy knowing that community knows things better than they used to, so to speak. When we do those sessions and you see the light bulbs go off, which I think is quite fun. Even to get back to that weeds brochure we did. And people have been in the valley for a while. I was relatively new. I was quite surprised that they didn't know what was on their land.

The group has produced a 14-page brochure titled 'Weeds of the Upper Kangaroo Valley', which it regularly shares with the community. The weeds it describes include small-leaf privet, lantana (*Lantana camara*), blackberry (*Rubus fruticosus*), cassia (*Senna pendula* var. *glabrata*), wild tobacco, Japanese honeysuckle (*Lonicera japonica*), Madeira vine, cape ivy (*Delairea odorata*), climbing groundsel (*Senecio angulatus*), English ivy (*Hedera helix*), fireweed (*Senecio madagascariensis*), Parramatta grass (*Sporobolus africanus*) and giant Parramatta grass (*Sporobolus fertilis*). For each, photos are included as well as a description of the plant, how it is dispersed, how to distinguish the weed from look-alikes and the most effective control methods.

The message of the lyrebird

Spring was in the air one night in early September 2022. A warm breeze cut through the remnants of a winter chill in Upper Kangaroo River. As I arrived with my husband and 8-year-old daughter, there was a growing sense of excitement around the Upper River Hall as car after car, including many an electric vehicle, slowed down to find a spot to park.

My daughter was thrilled at the sight of the suspension footbridge. It beckoned to her, calling for her to run across it then stop suddenly halfway in the hope of spying an elusive platypus, although her thunderous run across the bridge would surely have sent any diving.

But even the bridge or the prospect of platypuses couldn't hold her attention for long. Her passion for popcorn soon had us joining the steady stream of people wandering over to the hall. And she was not disappointed. Not only were there refillable containers of popcorn but an esky full of ice-creams. Better still, there was a mature red cedar tree, just beginning to bud, begging to be climbed.

Slowly, beneath her, the outside area filled with people. Chatting around the small fire pits, eagerly awaiting the start of the documentary. I even bumped into someone I knew who I used to play footy with in Sydney. The ties to Sydney are strong in Kangaroo Valley.

There were of course familiar faces from the various Landcare groups I had come to know in the area.

The hall was set up like a cinema. Rows of plastic seats were lined up, scattered with a smattering of eclectic cushions. At the front, a screen dropped from the ceiling with the distinctive green Landcare logo and 'Upper Kangaroo Valley Landcare Group Inc.' in large white letters written alongside it.

Eventually the crowd made their way into the hall until all the available seats were taken. Kids of various ages sat on cushions up the front. The official attendance count was 91.

Before the proceedings began, a member of the Landcare group gave an Acknowledgment of Country and explained what Upper Kangaroo Valley Landcare does. Here was a captive audience, and a chance at recruitment.

As if right on cue, the heavens opened and rain thundered down on the hall. The timing couldn't have been more perfect. The hall felt that little bit cosier.

Time for the movie to start.

This 80-minute film takes a deep dive into the lives of lyrebirds and their connections to Country. In doing so, it sets out to find out whether lyrebirds in the wild mimic human noises like the ones in captivity. It also asks deeper questions about the impacts that humans have on the planet as well as the way that nature responds to and recalls its interactions with people.

While the screening of the film was used as a fundraiser for Upper Kangaroo Valley Landcare, raising over $1,000 for group activities, it also reinforced the wider aims of the group: to educate the community about our diverse native ecosystems.

In reflecting on the night, my husband described it:

> A good entrée into the sort of work they [the Landcare group] did because a lot of the people who featured in the film were either enthusiasts or citizen scientists who were participating in studies of the lyrebird, and culminating in a citizen scientist effectively recording a [wild] lyrebird sounding like a person talking.

The film showed how everyday people can make a difference to our knowledge and understanding of the world around us.

Where neighbours become good friends

Beyond fundraising, a chance at recruitment and education, the movie night clearly contributed to an unwritten and often unspoken aim of the group: to promote community – indeed only one of the members interviewed mentioned it. Despite not being a formally recognised aim of the group, promoting community is something that many members value; it provides a reason for joining and keeps them coming back over the years.

> I totally joined to get to meet the community because I was new ... I moved in when I didn't really know anybody in the area. So, it's nice for me doing kind of parallel playing like kids, it's like I'm doing something, I've got something in common with this person, it's easy to chat to them, because we've got something

> in common. It's not an intense experience, it's not like a dinner party or something like that. It's just going and having a very casual sort of thing, where you catch up with lots of different people as well.

It is the ability to connect with people in the community, regardless of their ability to control weeds that is ultimately valued.

> Everybody's supported and the labour is valued. Even if - you know, some of our members are quite old and quite frail, but they turn up with biscuits ... or chat ... if you're physically able, you do weeding, but that's not the be-all and end-all of it. It's actually Landcare, you can contribute in a variety of ways ... that's a good example of a successful organisation. Someone who can't weed but still would attend. That's pretty good.

The Landcare lifeline

The importance of the group has never been more vital. In 2019–2020, Landcare connections were one of the keys to the community supporting one another through the threat and impact of the Black Summer bushfires. The year before, the Upper Kangaroo River Progress Association convened the community and established nine Bushfire-Ready Neighbourhood Groups[15] that would support one another to prepare for, and in case of, a bushfire. While the bushfire-ready groups extended beyond the network of Landcare members and the area of the Landcare group, the relationships built through Landcare were important.

> You could talk to other people in the Landcare group about what they were doing, what their [bushfire] plans were, we swapped plans ... we shared similar kind of knowledge about what to do. It was a good structure ... basically, the Landcare structure.

Then, in 2022, Kangaroo Valley and Upper River received almost a year's worth of rain in just four days.[16] This wreaked havoc on multiple roads into and out of the valley. You could no longer access Kangaroo Valley via the road from the north. The road from the south was barely passable, vehicles needed to wait in long convoys from Nowra to traverse the single lane that remained open. The roads from Berry in the east also experienced landslips. The damage was so extensive that all the roads were still undergoing repairs over a year later.

In the middle of the valley, the Upper Kangaroo River Road became cut off by a landslip. Thankfully there was some advance notice, so some residents managed to park a car on the other side of The Slip, before it became closed to vehicular traffic. But for those who didn't, or couldn't, other residents banded together to help them. For weeks they brought supplies, helped clear fallen trees[17] and provided carpools for those without a car 'on the other side'.

For the Landcare group, The Slip posed an additional challenge. Not only did it make it hard for those downstream to join Landcare activities but the sheer volume of rainfall the area received gave weeds a boost, making Landcare activities more important than ever.

The first time I joined Upper Kangaroo Valley Landcare, in May 2022, The Slip had only just re-opened. By the time of my second visit, in June 2022, it had closed again. And so it went, opening and closing over the course of 2022 and 2023. Yet the group was determined to meet. And not just to keep on top of the weeds. Each time The Slip closed, residents were just as, if not more, isolated as they had been during the COVID-19 pandemic lockdowns. Landcare provided a much-needed chance for residents to connect and support one another.

> Landcare I'd say is embedded in the Upper River community ... It's a way of knowing everybody, so obviously for social reasons that's really, really important, but also ... in rural or regional areas, it's more important to know who your neighbours are because of floods and fires ... you're far more aware of how you rely on networks, because services in the regions are not available.

Growing social capital

Through all the accounts of the Upper Kangaroo Valley Landcare and Bushcare, it is evident that they have been highly proactive in engaging with a range of government and non-government organisations. Of note is their long-term relationship with Commonwealth Scientific and Industrial Research Organisation (CSIRO) scientists who research biological controls for the management of weeds. In 2011, a property near Gerringong Creek Road was the site of a CSIRO release of a 'white-smut fungus ... to control the serious environmental weed mistflower'. It was remarkably successful in reducing the cover of this weed in the valley. The next year, leaf-eating beetles for control of Madeira vine were scheduled for release alongside Gerringong Creek.

Again, in late 2019, when CSIRO was ready to trial the release of a fungus for control of wandering trad (*Tradescantia fluminensis*), Upper Kangaroo River was one of the first places it was released and monitored as a nursery site. This required the group to stop controlling the weed through other means so that CSIRO could monitor the effectiveness of the release. The group was excited to be part of this programme because it meant that they could deal with an intractable weed in a sensitive riparian ecosystem without the need for widespread application of herbicides.

Beyond their relationship with CSIRO, the group has built relationships with many other organisations within and beyond the region. These include the Shoalhaven City Council, Sydney Catchment Authority, Southern Rivers Catchment Management Authority, South East Local Land Services, NSW National Parks and Wildlife Service, NSW Department of Primary Industries and Regional Development, Green Corps and Conservation Volunteers Australia, among others. The engagement with these diverse organisations reflects the group's commitment to ecological restoration of Upper Kangaroo River.

Upper Kangaroo Valley Landcare also works closely with other Landcare groups in the area, most often the Kangaroo Valley Environment Group and Brogers Creek Landcare. Together, these groups have applied for multiple grants over the years for the control of invasive plants and animals. In this way, the group not only enhances the ecology of Upper Kangaroo River but also contributes to environmental collective action across Kangaroo Valley.

Where to now?

A year after I first joined the group, I could almost be considered an active member. I had been to multiple working bees and the movie night. Being the closest of the groups to Sydney, and working bees mostly being on a Monday, I found it easier to get along to activities than the other groups.

The second year there was something different about going along to the working bee that preceded the Annual General Meeting. Instead of looking forward to something new, I was looking forward to something familiar.

I knew that as I arrived, a jolly and fun group of six or seven people would emerge from their vehicles for some weeding in a place that had local significance: the place where the group had started its work almost 30 years ago and that had been transformed by these efforts.

I looked forward to an hour of weeding alongside passionate Landcare members. I knew I would laugh; I would be entertained by their stories and antics.

I also was sure that I would learn something new, about the plants, or the river, or the animals. Last year I had learned that bandicoots lived in the area and discovered what their scratchings looked like.

I knew to carry insect repellent with me – I'd managed to pick up a tick at the movie night. So much for weeding being the risky work!

I looked forward to stopping weeding earlier than usual to celebrate not only the achievements of the morning but of the preceding year.

I knew there would be an abundance of homemade cakes to indulge in. While usually I would bring along a cake from the local bakery, this time I wanted to show how much I appreciated the group by baking a cake – a Spanish Torta de Santiago.[18]

True to form, the morning played out just as I anticipated. We pulled out moth vine and Madeira vine. We laughed and ate cakes, brownies, Portuguese tarts and macaroons. We celebrated the 200+ hours of volunteer time spent looking after the valley in the preceding year. And plans were made for where working bees would be undertaken for the rest of the year. This time the names of the meeting spots were familiar to me too. I looked forward to the next time I could make my way to Upper Kangaroo River and join in the group activities.

The one thing that Upper Kangaroo Valley Landcare needs to keep this momentum going are new members to participate in working bees and take on administrative roles. To revitalise the group, it may be time to get back to basics. Perhaps it is time to rethink what weeds the group targets or where the group focuses its efforts.

It took Towamba Valley Landcare (Chapter 9) and Deua Rivercare (Chapter 6) crisis meetings to draw attention to the future of their groups and inspire younger and newer members to step forward to take the lead.

While Upper Kangaroo Valley Landcare did advertise that the 2025 Annual General Meeting would involve a discussion of winding up of the group, it may need a special event to draw attention to the challenges the group of its broader membership base. Making a different time and place to recognise the achievements of the group while also discussing the reality of its struggles may mean something new emerges. While it might be daunting to do this, at least it would provide the members and those who support the group with an opportunity to celebrate the 30 years of achievements and the sense that they have done all they can to keep it going.

Beyond that, there are others in the broader Kangaroo Valley who are interested in starting a Landcare group. There may be opportunities to give such a group a head start by sharing three decades of lessons from the group, paving an easier path for others to follow.

5

Restoring habitat in Tapitallee and Budgong

They lost everything in the Currowan fire, aside from the house. Twenty-three hectares of property burnt to a crisp. Not a skerrick of green was left.

With the house mostly unscathed, what was most upsetting was the eight-and-a-half years of work that had gone up in flames in just 10 minutes. Years of dedicated labour, crawling in among the woody weeds, cutting and painting them with herbicides, coming out covered in leeches. Despite the challenges, they had persisted in slowly, methodically removing countless weeds: blackberries (*Rubus fruticosus*), lantana (*Lantana camara*), cassia (*Senna pendula* var. *glabrata*), fireweed (*Senecio madagascariensis*) and coral trees (*Erythrina x sykesii*).

And then there was the time spent planting trees. One thousand trees and their 1.8-m-tall chicken wire cages, purpose built to prevent damage from deer. Burnt. Gone.

It wasn't just a sense of loss at the scale of the work they had invested, it was the work of the Landcare group too. The neighbours who had come over to teach them about how best to control the weeds, give them advice on which were the most suitable trees to plant and then hours spent in Landcare working bees to help get the work done.

> They've all been set back because of the fires, so all the hard work that went into everything up until 2019 ... got all wiped out.

The Landcare members grieved for their friends; the loss was shared because the group had invested so much in this property.

Of course, it wasn't long after the fires that the Landcare group sprung into action again. The relationships built over those eight-and-a-half years were as strong as ever.

Members of the Landcare group came once a week, every week, to help clear the dead trees and the rubbish. Through the COVID-19 pandemic they kept coming, socially distanced, to clear the black wattle (*Acacia mearnsii*) that thrived after the fire, to make space to plant back trees. Together, they've planted 1,700 trees. In doing so, the group has continued to demonstrate their long-term commitment to rehabilitating private properties along Bengalee Creek.

> The basic aim [of Tapitallee Budgong Local Landcare] is to reduce the weed cover in the area and replace that with endemic species of plants that provide a habitat for biodiversity.

The birds and the trees

If you're from Sydney, you've no doubt heard about or visited Kangaroo Valley. It is well known for its rolling green hills, quaint village and striking sandstone bridge. Tucked away behind Kangaroo Valley are Bengalee, Budgong and Tapitallee, small communities nestled between the Great Eastern Escarpment and Bengalee Creek, a tributary of the Shoalhaven River. There are no townships to speak of, just a smattering of agricultural properties – dairies and wineries – and rural residential blocks.

The views looking up to the cliffs – the grand sandstone rocks that emerge through the Illawarra subtropical rainforest and keep watch over the valley – will take your breath away. One Landcare member recalls the day they first came to the area:

> The day that we came, it was a Sunday afternoon, and it was chucking it down with rain. We couldn't see from halfway up the

> escarpment, and then all of a sudden it stopped, the clouds lifted, and there were waterfalls just coming off the cliffs.

Nearby is the Bundanon Art Museum, built where Arthur Boyd – one of Australia's pre-eminent modern artists – set up a studio in 1982. The surrounding environment – the river and the bush – provided inspiration for many of his paintings and pottery.

If you've been to Parliament House in Canberra or the Arts Centre Melbourne, you may be more familiar with this landscape than you realise. The Great Hall Tapestry, painted by Boyd and then stitched by 12 weavers, evocatively captures a scene of towering eucalypts as viewed from Bundanon. Fourteen oil paintings in the foyer of the Arts Centre also portray the mesmerising trees, river, rocks and sky of this special Country; each artwork 'a homage to the majesty of the bush'.[1]

What is less frequently captured in Boyd's artwork, or the description of the escarpment above, is the diverse birdlife that benefits from the abundance of native vegetation. The tapestry only features a lone sulphur-crested cockatoo (*Cacatua galerita*) and the 14 landscape paintings only include two black cockatoos and a swan.

BirdLife Shoalhaven in contrast have recorded 400 bird species in the region, at least 100 of which have been recorded by Landcare members on their properties. As one Landcare member explained:

> So, bassian thrushes are scratching around, and the lyrebirds have been very active ... We might hear the tree creepers, we'll hear the gerygones, the golden whistlers were down there yesterday. Lewin's [honeyeaters] are down all the time. Lots of yellow-faced honeyeaters passing at the moment. The bowerbirds are around, and of course the crested shrike-tits are around at the moment, the red-browed finches, the whip birds, scrub wrens, the little fairy wrens are around ... Robins, they're here all the year round. The little brown thornbills are here. The striated thornbills were here this morning, and the spinebills and the grey fantails are here ... And the wonga pigeons are still here.

Beyond the birds listed by Landcare members, there are also endangered and special birds in the area, including the glossy black cockatoo (*Calyptorhynchus lathami*) and the rockwarbler (*Origma solitaria*), NSW's only endemic bird.[2] The nearby Tapitallee Nature Reserve supports vulnerable populations of the powerful owl (*Ninox strenua*), gang-gang cockatoo (*Callocephalon fimbriatum*) and glossy black cockatoo.[3] The Bugong National Park provides habitat for the threatened little lorikeet (*Glossopsitta pusilla*). The 43 bird species formally documented in the nature reserve and 63 documented in Bugong National Park are about half of those documented on nearby private properties. These numbers speak to the interconnections between and ecological significance of the properties whose management is assisted by the Tapitallee Budgong Local Landcare.

Indeed, birdlife in this area has been highly valued and long documented, well before Tapitallee Budgong Local Landcare and BirdLife Shoalhaven came into being. From 1926 to 1940, one of the residents, Aubrey Elliot, lived where Tapitallee Budgong Local Landcare now cares for land. He kept detailed notes and photographs of birds in the local area and interviewed other residents to collect information on the status of birds. He recorded 153 species, 19 of which have not been recorded in recent observations.[4] Thirty-five additional species have been recorded by Garry Daly, an ecologist who lives in the area.[5]

It is important to paint this picture (pun intended) of the culturally and ecologically significant trees[6] and birds because they inspire Tapitallee Budgong Local Landcare to work together on weeds. Conversations with and among the group invariably move from trees and birds to weeds and back again. These three elements are so closely related that it is hard to tease them apart, as this quotation shows:

> I went down here a couple of mornings this week, I came down because I'm keeping an eye on the lyrebird and I wanted to go down and slash that track out [of lantana] a bit. Early in the morning, just after sunrise, is when they're most active. But yes, there's definitely more birds around. And stopping, like not

> just flying by. There's a little grey fantail. You can hear him. Bip. Bip. But there's not many at the moment, but there was a lot. That's a kurrajong [tree]. They're nice. We've planted a few of them here. That's another trema [orientalis]. Oh, that trema [tree], the silvereye were in that one the other morning. See that bushy one? So they've been really good for the birds already.

While all the members of the Landcare group are passionate about the environment, it was the weeds, and the lantana in particular, that brought them together.

Learning about lantana

The beauty of the area attracts rural retirees, some of whom are 'blissfully ignorant' of the weeds. The native vegetation, that is so striking at a distance, obscures an untold number of weeds. Unsuspecting new residents think of the multi-coloured lantana flower clusters as 'cute and pretty' and fireweed as a plant that has 'beautiful yellow flowers'. Often it takes neighbours or others in the know to point out weeds to the new residents.

> Some time ago, I had a friend who is a member of the Australian Natives Association ... And she came onto the property and as soon as she got here, she pointed out Parramatta grass, which I hadn't known about ... So my knowledge about, and my reactions to weeds, has very much been a function of what has been pointed out to me.

Slowly but surely, the new residents come to learn about the wide range of weeds that have crept or are creeping into the native vegetation. They talk about woody weeds, the 'stands of lantana', the 'big [wild] tobacco bushes' and the 'solid cassia'. Then there are the vines, the Madeira vine (*Anredera cordifolia*) and particularly the moth vine (*Araujia sericifera*), as well as the 'carpet of yellow' fireweed and the Formosa lily (*Lilium formosanum*). So while lantana is a key focus of the group, it is only one of many plants of concern to members.

But Landcare members rarely learn about the plants on their own. Neighbours are instrumental. One of the advantages that Tapitallee Budgong Local Landcare has is their access to experts who live locally. For example, Garry Daly is a zoologist who has been documenting and writing about the plants and animals of the Shoalhaven for over 30 years and whose 'knowledge is encyclopaedic'. Such experts share their expertise with the group, not only about the control of weeds and pest animals but also about the native vegetation and what it needs to aid rehabilitation. Members and non-members alike value the contributions of these individuals.

> They do understand a lot more about what's going to happen in the long term, and how they can stop seed banks moving in, and manage deer properly ... Garry Daly, as I said, he's a zoologist. He knows everything there is to know. If I'm not sure, I'll text him a photo, he comes back to me straight away with what it is. And it's having that resource.

In talking to neighbours, learning about weeds and watching the landscape change, residents come to notice that if left to their own devices, weeds quickly spread and begin to dominate, 'smothering native vegetation and toxic to soil'.[7] One resident described lantana as 'a weed that is most concerning because it's a functional ecosystem driver':

> It's gone through several phases over 35 years, and changed, became, a lot of wattle landscape, and then as the wattle senesced it became lantana and wattle, and then some areas just became, well, the north-facing slopes became basically just stands of lantana.

Of all the weeds described, lantana is the most concerning to members because it forms 'dense thickets'.

> The centre part up the creek was just solid, and I mean solid, lantana. Impenetrable. High, up into the trees. Forty feet up into the trees. It was just huge.

On another property, one of the members explained: 'And this, I call it the Top Triangle, was all lantana. So much that you had to cut your way through.'

These descriptions of lantana are consistent with scientific descriptions of the plant, which refer to it forming 'dense impenetrable thickets'[8] and one of Australia's 'most debilitating invasive weeds'.[9] Lantana is listed as a Weed of National Significance because of its potential to spread easily into agricultural land, native ecosystems and urban areas. Indeed, it is not only problematic in Australia; nearly 50 countries have declared lantana a weed.[10]

Historical descriptions of lantana are similarly evocative. An article in the *Shoalhaven Telegraph* in August 1888[11] reported on the spread of weeds in the Richmond River district, in northern NSW, ~900 km away. This is the first mention of lantana in the local newspapers:

> The lantana has taken entire possession of some portions of the country, and is a fearful pest. On each side of the road from Wyrallah to Lismore, and also for some miles around Lismore, hundreds of acres of land that had once been cleared have become utterly useless through the spread of this shrub.

Only one month later, the same newspaper reported:

> Alderman Monaghan thought the Inspector of Nuisances must be asleep, as he did not appear to notice that a very noxious weed, 'Lantana' was beginning to obtain a footing in the municipality.[12]

In 1905, seven years after these articles were published, Nowra decided not to formally declare lantana a noxious weed in the area. Nearby Berry Council formally declared lantana a noxious weed in 1907 and Jamberoo Council followed suit over a decade later in 1918. In 1926, despite multiple concerns raised about lantana, Nowra Council continued to hold out against declaring lantana a noxious weed while Shellharbour City Council finally made the declaration.

Finally, in 1932, almost 90 years after lantana was introduced to NSW, 44 years after a local councillor raised concerns about lantana in the municipality, and 27 years after the first formal discussion was had about whether to declare lantana a noxious weed, the motion was carried in the municipality of Nowra.[13]

Beginnings of Tapitallee Budgong Local Landcare

How do you tackle something so overwhelming? You start small. Tapitallee Budgong Local Landcare began because of the efforts of one landholder, who obtained a grant to fence off and control 8 ha of lantana along the creek lines on his property and begin a process of revegetation. Even from the outset this landholder knew he wasn't on his own; a neighbour was instrumental in alerting him to the grant and was also passionate about controlling lantana on their property.

Very soon it became obvious that efforts on one property would be futile if neighbours didn't control lantana too. Lantana, like so many weeds, presents a collective action problem because of how easily it spreads across property boundaries. In the case of lantana, birds are one of the vectors that helps the plant spread.[14]

On the advice of Landcare, the landholder agreed to host a workshop on lantana spraying in 2009. This not only provided him with an opportunity to meet with his neighbours and raise his concerns about lantana but it provided an opportunity to test how much interest there was in working on the problem together. Soon thereafter Bengalee Landcare (as it was known then) was formed.

In the early years, the group mostly worked under the umbrella of other Landcare groups and took advantage of funding sourced by the Shoalhaven Landcare Association. Initial funding was for individual landholders to control lantana on their properties, largely through the use of herbicides. Funding was also available for purchasing materials and tube stock to facilitate revegetation efforts. Through such projects, members of the group became familiar with the best strategies for controlling lantana. They were able to experiment with a range of equipment available from the Shoalhaven Landcare Association and learnt first-hand about the challenges of getting neighbours to control lantana in the short and long term.

These projects helped 20 to 40 landholders in the region but not all of them were members of Landcare. In providing such support, the grants helped to identify who was most passionate about controlling lantana and who had the capacity and interest to commit to controlling it over the long term. The group learnt about the importance of getting landholder commitment in writing, as this increased the likelihood that neighbours would deliver on promised weed control activities. These activities resulted in a group of five to six properties emerging as core members.

It wasn't really until Tapitallee Budgong Local Landcare received its own grants in 2018 from the South East Local Land Services and the Shoalhaven Landcare Association that the group really settled down into a regular pattern of working bees, newsletters and social events. The Bengalee Lantana Clearing Project aimed to clear 8 ha of lantana from the southern Illawarra escarpment at Tapitallee.[15] The project spanned three adjoining properties and linked together revegetation efforts that had previously been undertaken on two of the properties.

To make the project manageable, the group divided the 8 ha area into nine zones, with a clear management plan for dealing with the lantana in those areas. The project began by getting a contractor with an excavator to clear tracks through the dense thickets. This then enabled members of the group to control the lantana by volume spraying, hand spraying, cutting and painting and hand pulling. Once the lantana had died off, the dead canes were mulched or slashed to clear the treated areas for planting. While it was back-breaking work, the group also got creative with methods of removing the lantana.

> Member 1: It's fun doing it as a group. As much as it is hard work. People sweat and bleed and all sorts of things!
> Member 2: Fall down the mountain and do forward rolls! 80-year-olds!
> Member 3: In Zone 10 we rolled the lantana, so the guys had big poles, we took big poles down so we used to take the poles and roll it. Pull the roots out and roll and roll and roll. And one ended up down in a hole! And so did another!

> Member 4: Yeah! You get some momentum, and over you go! You follow it down the hill!

Similar to other groups, like Upper Kangaroo Valley Landcare (Chapter 4) and Tathra Land Care (Chapter 7), that clearly define the boundaries and sub-boundaries of where and when they will clear weeds, Tapitallee Budgong Local Landcare created zones within their 8 ha area to help with the tree planting part of the project. They divided the area into 10 zones and planted over 800 trees across them, carefully documenting what trees were planted in which zone.

This project provided the group with an opportunity to promote their work to landholders in the wider region and invite them to be involved. An article in the *South Coast Register*, a local newspaper, in April 2019[16] 'invites everyone to a working bee'. The call was clearly successful because a follow-up article reported that 26 volunteers worked to plant 412 trees and invited people to join the next planting day. A similar number of people attended the public inspection days.

Five articles about group activities published in the *South Coast Register* in 2019 were written as educational pieces, designed to explain to the broader community the amount of work required to prepare the sites ahead of tree planting, the extent of lantana and why it is problematic, the multiple types of labour required to remove lantana and the need to create sturdy tree guards that can withstand wombats, wallabies and deer. The articles explain how tree species are selected and the benefits that tree planting should have on reducing lantana regrowth, as well as providing habitat for native species. They also promote the benefits of working together and the sense of community built.

From their past work, the group knew that these activities were just the beginning. The removal of lantana and planting of trees that took place over 18 months by Landcare members and the community would require years of maintenance, including regular watering and lantana removal, as it would invariably regrow. The zones facilitated follow-up work, documenting watering efforts and areas needing lantana regrowth control.

> I say to people, a way to get rid of lantana is persistence and it means persistence over a number of years. And you, possibly, if you've got lantana, rather, you never actually get rid of it, because the birds actually spread it from a distance away.

This cross-property activity has formed the foundation of work undertaken since then, demarcating priority areas, as well as cementing relationships among landholders. Monthly working bees, central to the lantana clearing project, have become the backbone of the maintenance work. The spirit of volunteering, with 975 hours contributed to the 18-month project compared to 39 hours of contractor time,[17] also continues apace.

A weed in time saves nine

The philosophy of Tapitallee Budgong Local Landcare is that in order to reduce lantana's foothold in the foothills, they need to plant trees because 'lantana doesn't grow well in shade'. This is a long-term strategy, which requires the group to control regrowing lantana regularly until the vegetation they have planted is tall enough to keep the weed at bay.

Initially, they expected that the lantana clearing project would be the main outlay of time and effort – 'the grunt work' – required to get lantana to a manageable level, and then property owners would continue to control lantana on their own properties, with some assistance from the Landcare group through working bees. This approach has been more or less effective.

It has been more effective on the properties of landholders who are active members of the group. In part this is because these members are proactive about controlling lantana in their own time: 'I'd probably spend 3 to 4 hours a week just in maintenance. It's a full-time job. Otherwise, it just gets away.' Similarly, another member explained:

> I now spend around 4 hours every week manually grubbing out new lantana growth including the roots. My experience is that

> unless constant attention is paid to the problem the weed will rapidly regrow.

The benefits of this ongoing control is evident to others who work with the group, who refer to these properties as being 'really, really good quality' or 'perfect'.

But even these dedicated members find it difficult to make time to control weeds and water trees, if someone becomes sick or is injured.

> Member 1: [they were] quite an active member, got ill, sick ... for the last 18 months.
> Member 2: They hadn't been doing their maintenance ... during the drought we probably lost 30 per cent of what we planted.

So maintenance is not only important in terms of weed control but also in terms of ensuring the success of tree planting efforts.

Another challenge for the group in terms of maintenance is associated with changes in land ownership. While some of those involved in the initial project were keen and enthusiastic, the landholders who have since taken over some of those properties are less engaged.

> [He] has been to a couple of our meetings, and he's been to a working bee, but he's also working and really busy, so this is, as you can see, it's a huge problem, so you've kind of got to be on to it all the time. And I think, he's away at the moment.

The impact of land ownership changes is also evident to those who work with the group. They can see where past investment of effort has been undone because 'the new landowner is not remotely interested ... He's supposed to be doing his own contribution and all he does is slash the track and that's about it.' People external to the group also recognise that while the Landcare group has a strategic plan for the area, not everyone is equally committed to ensuring that their part of the plan is met. This means new weed seeds continue to spread.

Together, the combination of sick members, changes in land ownership and landholders who are not committed to long-term maintenance means that the group has watched some of their hard work disappear under new thickets of lantana.

The challenges of changing land ownership

We made our way up the hill, across the paddock. Once we were through the gate, the grassland came to an abrupt end. There were trees everywhere, and at first sight, I couldn't spot any lantana, even though I was reliably told that when the group had started, this was 'completely lantana'. Now, the turpentines were thriving, easy to spot with the chicken wire tree guards around them, still protecting them from deer.

The Landcare members, whose eyes were better trained than mine at spotting regrowing lantana, were quick to point out where they could see lantana making a comeback. Here the lantana seemed manageable, the plants were few and far between, easily pulled out by hand.

But as we came to the boundary fence, the lantana was thick. It was easily as tall as us, if not taller. This lantana, they explained, was treated at the same time. It had all been cleared in 2018, but maintenance had been an issue for the past owner.

We went through a gate in search of the trees that had previously been planted here. It was like being at a kids' birthday party and playing hide and seek. The trees were very well hidden, and even with four of us on the hunt pushing our way through the grove of lantana, it took a while before we could find them.

In the end we found a handful of trees, but they were in the fight of their lives, trying to keep their heads above the lantana. The group looked on, dismayed. Without a machete, there was no way we could continue searching. We were left wondering whether the new landholder would come to the rescue of the trees or whether this would become another case of lost effort.

Thankfully, hope wasn't far away. We visited another property. This one more regularly maintained. Here the lantana was mostly under control and the new trees were thriving, towering well above their 1.8-m-high tree cages. Handwritten labels were attached, providing details of the scientific and common name of each plant.

The Landcare members also pointed out trees that were coming up on their own. Sandpaper figs (*Ficus coronata*), giant stinging trees (*Dendrocnide excelsa*) and red cedars (*Toona ciliata*). Each tree a testament to the value of their efforts.

But rewards like this don't come without commitment. The group discussed the way that they not only had to come back to clear the paths but also carefully tend to the weeds growing in among the trees and the cages.

> Member 3: When we had our working bee, everybody helped us to get rid of the regrowth lantana and weeds inside the cages as well. Because a lot of – inside the cages, the lantana has come up inside the cage!
> Member 4: The cages have little wires on them. We take them off and weed around the plant and then put the cage back on.

Beyond seeing the trees grow, the members have noticed the birdlife change. While some members were initially concerned that removing lantana would have an impact on smaller birds that used it for habitat, they have been relieved to see that after some initial declines in those populations, they have since returned. There are also many more species that seem to be thriving in the cared-for Country.

> But certainly the work I've done has seen an increase in some of our more rare species of rainforest birds, like logrunners. We're one of the properties with logrunners on them this far south, and I've seen that population expand over the 20 years I've been here.

The group's newsletters also report on the benefits of past projects, revealing just how quickly things can change.

> It will be 3 years in April since we planted our first trees. We observed Lewin's honeyeaters feeding on the ripe berries of the native peach for the first time last week. (March 2022 Newsletter)

Others who work with the group have also noticed a change in native animals:

> I see so much change in the wildlife. Stuff you just wouldn't see coming out in the open. Frogs, bush rats, swamp wallabies, owls. All that out in the open. Take the lantana out, put the good stuff in, get rid of the foxes.

Those external to the group note the benefits that collective lantana management has provided, not only to native plants and animals but also during post-bushfire recovery.

Weeding black wattle

The fires have forced Tapitallee Budgong Local Landcare, along with countless other fire affected properties on the South Coast of NSW, to face large-scale ecological change and critically reflect on what plants are considered weeds. Talking to landholders from Kangaroo Valley all the way to the Victorian border, there is a similar story told by those whose properties were burnt.

First, residents had to confront what was left after the fires. The photos depict a sea of ash, from which black trunks stood like sentinels and burnt-out cars left no evidence of what colour they had been. Even the pools of metal from the tyres were only recognisable as tyres because of their location near the car frames. They simply appeared to be abstract silver-coloured artwork.

Then some colour began to re-emerge, but not consistently everywhere. It was clear the fire had burnt hotter in some places. Even two years on, those places were still taking longer to recover. Where the land was awash with green, it was obvious that with the subsequent wet years, the weeds were thriving. And so too did many natives. But the way some native plants thrived in the post-fire environment raised questions about what was a weed. Plants that caused the fires to burn hotter were now being critically re-examined.

Before the fires, it was mostly non-native plants, like lantana, privet, camphor laurel (*Cinnamomum camphora*) and running bamboo (*Phyllostachys* species) that were a specific concern to organisations like the NSW Rural Fire Service because they 'can contribute to the intensity of bush fires'.[18]

Since the fires, many landholders are concerned about the thickets of black wattle (*Acacia mearnsii*) that have emerged because they observed that places burned especially hot where stands of black wattle were dense. While landholders are aware that black wattle will senesce over 10 to 15 years, they are concerned that the next lot of bushfires will come through before then and will be exacerbated by the plant.

> Member 2: We've had a couple of properties that were burnt out in the bushfires, and one of our members along their creekline, it's just – well, we've done a couple of days, haven't we, and it's almost impenetrable black wattle over your head.
> Member 4: Yeah. Just like that.
> Member 2: Just in 12 months. And it was all, you know, well, it was all burnt in the bushfire, so that's 2019, 2020. And it's actually become a weed. Even though it's a native, it was a weed, because it's just choking up the whole – and choking things out. You probably could wait it out, and it would die off, but ...
> Member 4: But it's creating additional fire load. So, it's almost overwhelming.

Others associated with the group have referred to black wattle as a 'fuel bomb', and even contractors and council staff who work in the area treat black wattle like a weed. Not only do they see it as a fire hazard but that it also creates environmental conditions that encourage lantana growth.

> It's like lantana. It stops anything native growing through, same as privet. It lasts all of 10 years, then it promptly dies, and knocks down anything that's trying to struggle through it, and then lantana takes over as the primary responder ... It is definitely helping lantana. Overwhelmingly helping lantana. So, my policy now is, black wattle goes. I treat that like privet. It's a weed.

Driving around properties that are recovering after being burnt, it is easy to see why they are concerned about the black wattle. On the other side of the fences, where their neighbours have not been controlling black wattle, in just two-and-a-half years, the wattle has grown to 3 m. Like the lantana, it is dense, and it is hard to imagine any other vegetation making its way through.

Others in the community don't agree that black wattle is a weed and are not committed to its removal, however they understand why the plant makes their neighbours anxious and appreciate the mental health reasons used to justify its removal.

> [Black wattle is] part of the ecological long-term function of systems that they have these periods of high regrowth. It benefits a whole range of mammals that don't get a chance to bounce back so quickly if that regrowth is not there

Nearby is Bundanon Art Museum and Boyd Collection Gallery.[19] There too black wattle is considered a 'native invasive weed' and work is undertaken to control it across the 1,000 ha property.

Bundanon and Landcare

While the focus of Tapitallee Budgong Local Landcare is on a handful of properties between the escarpment and Bengalee Creek, members of the group are passionate about rehabilitating the broader area. They have been instrumental to several projects undertaken at Bundanon.

Much of the 1,000 ha of Bundanon is:

> natural bushland ... about 80 per cent of that is in quite good condition. The remaining 20 per cent of that bushland in some cases is in very poor condition. Largely weed infestations.

The areas that are in the worst condition are those that were extensively grazed until the mid-1980s and then left largely unmanaged since then. This is despite there being a requirement as part of the original lease for Bundanon and other local properties to control weeds, including lantana, since at least 1924 (*Government Gazette of the State of NSW* 31 October 1924, p. 4997).

While Bundanon attracts many visitors – 69,285 in the 2023/24 financial year – only a small number engage in tree planting efforts.

Bundanon has had a decade-long partnership with Landcare Australia,[20] which has focused on land management stewardship.[21] While this partnership exists at an organisational level, in practice, it is groups like Tapitallee Budgong Local Landcare that provide regular and sustained support to enhance the vegetation and wildlife of Bundanon.

> It's a showcase now, so they've got to make it look good because it's a major gallery, regional gallery, basically. So, they can't have people and dignitaries coming there and they drive through a sea of weeds to see art and have a cup of coffee.

The benefit of working with Bundanon is that it enables the Landcare group to extend its project activities. Bundanon contributes to reducing the load of lantana in the region by drawing in volunteers from further afield, such as from the Albatross Navy Base, Conservation Volunteers Australia, Volunteers Illawarra and the Green Army.

The benefit of having such partnerships is that Bundanon is able to get assistance with 'primary weeding' as well as 'secondary weeding' to make sure that the weeds are kept at bay, enhancing revegetation efforts. Like Tapitallee Budgong Local Landcare, Bundanon subscribes to the view that revegetation is a long-term strategy, not only for the biodiversity benefits it provides but also to keep control of weeds. They are committed to long-term weed control and maintenance.

> Any weed site in bushland, in my opinion, you've got to continue to be able to maintain for 20 to 30 years before you can possibly think that you didn't have to weed it anymore.

Creating social capital

Through all the activities described above, Tapitallee Budgong Local Landcare group has built a strong sense of community. Although the working bees in this group are structured differently to those of other groups, food is key to building strong relationships.

Working bees involve working for two hours, from 8:30 a.m. to 10:30 a.m. on the third Saturday of the month. They take a half-hour break for morning tea before working through to lunch. It is usually the responsibility of the person hosting the working bee to provide tea, coffee and cakes. Although the group is careful to note that while food is appreciated, it is not essential.

> [Y]ou are working together, physically working together and you usually have a morning tea or something together, and it's a really – well, it's just a great way of getting to know your community and know where your support lies. When the fires came, look at the support we had. We wouldn't have had that if we didn't know all these people.

There are ~12 regular members among a membership base of ~41 individuals. This is a tightly knit group, who 'all care about the environment', relish being able to support one another with removing weeds but also enjoy one another's company. They, like other groups, speak of the importance of 'being part of a community of like-minded owners' and of the 'social cohesion and resilience where friendships are formed'. Indeed, for some members 'the community cohesion is much more important than the bush regeneration'.

> [M]y favourite thing about Landcare really, apart from the land care is the community it builds. It's a great thing that we have good neighbours and it's brought people together more. It's been wonderful. Some lovely social times.

It is clear from the interviews that there are a lot of overlaps between members with respect to what they like to do in their 'spare' time, when they are not weeding. This includes gardening, bushwalking, bird watching, playing tennis and pursuing artistic endeavours.

Perhaps it is these overlapping interests and friendships that drive the length of the working bees to be twice as long as most of the other groups, making time for one or two food breaks.

> We weeded around existing tree guards ... We also planted about 20 new tree stock saplings to replace those that had died and to supplement existing plantings ... We had a break for

> morning tea then braved the increasing heat for about another hour to finish our weeding and planting. This was followed by a welcome swim and a sumptious [sic] lunch, including the most delicious trifle - though much discussion was had of the relative merits of jelly in trifle![22]

This quotation comes from the regular newsletter, which is published more or less every month and is the prime mode of communication for the group. The newsletters include details of upcoming working bees and reports on past events, with photos documenting activities undertaken and progress made. They not only recognise the work of volunteers but pay tribute to the preparatory work done by landholders hosting each event and appreciation for food provided. In this way, they not only record work done but establish social norms that underpin group operations.

> We put out a newsletter, and so just to tell everyone in the broader area, you know, a hundred odd people that live in the broader area what we're doing and what we've achieved, and each time we have a meeting and, you know, it's a good record for us as to what we've done.

Beyond the working bees, newsletters include information about the organisation of the group, such as who is on the committee (comprising four regular members), volunteer positions that are vacant and other projects in which the group is involved. It also reports on activities and programmes run by Shoalhaven Landcare and provides updates on Landcare NSW and events.

One member spoke about the benefits of having such dedication to social activities and regular communication. They observed that over time, the social aspect of their group not only encouraged others to join but meant neighbouring landholders were more willing to hold Landcare events on their properties.

The importance of friendships was further evident during and after the 2019–2020 Black Summer bushfires. Landcare members not only knew one another but also each other's properties. They could see and feel the impact of the fires and were keen to lend a hand.

> But the social benefits of just getting together for a meal ... It's very much about listening to each other. But I think it's great. Anything that builds community is terrific, and then when you get the fires, you know each other better and you know each other's properties better too. That was another thing. When there's problems, you know.

Beyond Landcare, the relationships built among members of the group form the foundation of partnerships with key organisations such as BirdLife Shoalhaven which enable greater reach of their activities and support for their overarching goals.

> BLS [BirdLife Shoalhaven] and SLA [Shoalhaven Landcare] is a symbiotic and developing partnership. Many of its landowners are BLS [BirdLife Shoalhaven] members and supporters and know a lot about birds ... One of the best things we can do for bird conservation is to provide more habitat and reduce weed and pest pressure.[23]

What now?

In and around Bengalee Creek, Tapitallee Budgong Local Landcare provides fundamental social capital that underpins many environmental and bushfire recovery projects. By working together, members enhance native vegetation for birdlife, respond to drastic environmental changes and provide an emotional support network when disaster strikes.

The key question they face is whether they can encourage new neighbours to join their cause over the short and long term. Perhaps, like the other groups, they may benefit from diversifying some of their activities to draw new members in and demonstrate the many social and environmental reasons it is worthwhile to join a group like theirs.

INTERLUDE: ENGAGING AMENITY MIGRANTS

I can still remember the weekend my parents bought a rural block on a whim. My brother and I were away at Scout and Guide camps. When we came back on the Sunday afternoon, we could feel the electricity in the air. Our parents surprised us with the news that they had just bought a 10 ha property on the other side of the Blue Mountains.

The story goes that Dad was looking through the newspaper on the Saturday morning and saw some rural land for sale not too far from where our family loved to go bushwalking. It was a lovely day so my parents decided to go for a drive and check it out.

They met the real estate agent in the Kanimbla Valley and were shown several blocks of recently subdivided ex-grazing country. The moment they set foot on the block they fell in love with it. The hilltop had a smattering of gums and the views back onto the escarpment of the Blue Mountains were just sensational. They put an offer on it then and there.

A few weekends later they took us to see it. It was one of those days where the fog settled in low, all the way down to the creek at the bottom of the steep hill. Despite the lack of view, my brother and I could tell it was a special place. It has charm and a stillness that rings in your ears.

With the benefit of hindsight, I now see that my parents were oblivious to the weeds on the property. Not only were they both city slickers but they were also relatively recent migrants to Australia. They had grown up in Bogota, Colombia, and Birmingham, United Kingdom, and only moved to Australia 14 years earlier. What did they know about Australian plants? Next to nothing.

The place was covered in serrated tussock (*Nassella trichotoma*), with a healthy dose of blackberry (*Rubus fruticosus*) thrown in for good measure. Looking back, I recall that once a year, a paddock on the other side of the valley would turn bright purple. I have no doubt now that it was a paddock full of Patterson's curse (*Echium plantagineum*).

My parents were some of the first people to buy one of the subdivided blocks. They and all the other new landholders, mostly Sydneysiders, were blissfully unaware of the problem until a weeds officer from the local council came to tell them about it. Serrated tussock was one of the council's priority weeds, and they were keen to keep it in check.

Soon enough, the new landholders were getting on board with a joint aerial spray programme organised by the local council every few years. Given the soil, terrain and extent of serrated tussock, aerial spraying was seen to be the only economically viable way of controlling it. But there was one big problem: the largest landholder in the area wasn't interested in being involved. So eventually they gave up, and the serrated tussock continued to thrive.

What do amenity migrants know about weeds?

I tell my parent's story as a way to unpack the stereotypes that often apply to amenity migrants, also known as absentee landholders, lifestylers, tree-changers and hobby farmers. These terms are often used interchangeably to describe mostly affluent (sub)urban residents who purchase land or move to rural areas for specific lifestyle amenities, including the natural scenery, outdoor recreation or a sense of rurality.[1]

Amenity migration peaked during the COVID-19 pandemic, when thousands of city slickers in Australia[2] and elsewhere escaped to the regions, but this was simply an uptick in a long-term trend. Globally, there has been a notable urban-to-rural migration pattern, starting in the 1970s, that has fascinated human geographers and has captured the attention of long-term rural residents.

For decades, amenity migrants have been blamed for not doing enough to control weeds.[3] Their new and often part-time status in the community means farmers and government weed officers see them as people who are not present enough, don't know enough and don't have the skills needed to management weeds effectively.[4,5]

While these characterisations might have initially been true of my parents, and many other amenity migrants I have talked to over the years, I have often found these new rural landholders are keen to learn and make a positive contribution to their new community. This quotation from one of the Landcare members captures this enthusiasm well:

> I was blissfully ignorant. I thought that the little flowerets of the lantana were really cute and pretty and it wasn't until I learnt more about noxious weeds through my involvement with Landcare that we realised that we had a massive problem ... I made it my point when we purchased here to get in touch with any local community groups that had an interest in flora and fauna, so I got onto the Landcare website ... and immediately joined up and, before moving down here, made a point of travelling down on weekends where working bees were held by the group.

Recent research in south-east NSW undertaken by Nicholas Gill and colleagues[6] confirms this willingness to learn and act. They found that regardless of when amenity migrants learn about weeds, the overwhelming majority (97 per cent) now undertake weed control activities and are keen to do more.

Indeed, some of the weed officers I've interviewed over the years believe that amenity migrants are no more or less likely to control weeds than farmers:

> I'm careful to pigeonhole the absentees or the new landholders as the problem, because it's often not the problem. They come in with energy and enthusiasm to learn, and often it's easy to blame, but they're often not the problem. It can be old, traditional farmers.

The challenge of approaching new neighbours

What is evident from interviews with farmers and long-term residents is that part of the challenge of amenity migration comes from needing to form new relationships.

In part, the belief that amenity migrants have limited knowledge about weeds and different values and reasons for being on the land means that farmers are unsure about how to approach them about weeds. This reservation is compounded by social norms that discourage rural landholders from interfering with the decisions of others or causing tensions with neighbours.[7]

Weed officers also find it difficult to make time to engage with new landholders. Given that their usual working week is Monday to Friday, if they want to engage with part-time residents, they need to change their hours or work additional hours to meet new landholders on a weekend.[8]

Landcare provides a mechanism for overcoming these social challenges. For long-term residents, inviting new residents to join a Landcare group is much less intimidating and socially awkward than starting a conversation about weeds. For new rural residents, Landcare provides a place for them to start their journey of learning about the plants on their properties and how to manage them. With many Landcare groups meeting on weekends, they also provide information at a time and place that government staff can't provide on a regular basis.

Brogers Creek Landcare, for example, has made a special effort to make their working bees attractive to weekend residents. Having been founded by someone who was a long-term weekender, the group has always recognised that the timing of events and regular communication was important, as was the social dimension. Holding working bees on a Saturday afternoon gives weekend residents a chance to settle in before joining community activities. Being once a month, with a barbecue and cakes to follow, also provides a reason for part-time residents to make it to the valley on the Landcare working bee weekend. The timely email that comes out during the week before each working bee also makes attendance easy, with clear instructions provided on how to find and access the chosen property.

It takes time to build community

Thinking back to when my parents bought their property, the whole valley was one big subdivision. There were no long-term landholders living close enough to provide advice and no existing social groups for new residents to join.

Over the last 30 years, as more of the weekenders have moved to the valley permanently, my Dad included, the social relationships have flourished. There are annual get-togethers and regular social events at one another's houses. They have funded a community defibrillator stored in a central location as well as a community blackboard and book library. A Landcare group also took shape, planting many trees along the main road that connected the residents of the valley. Beyond participating in aerial spray programmes, the residents have banded together to control pest animals, particularly rabbits.

But in this community, there are no ex-farmers, people with agricultural backgrounds or horticultural skills to lead the charge on weeds. They have muddled through as best they can but the absence of pre-existing relationships in the valley is evident. There needs to be both the drive and vision, someone to identify the shared goal, as well as the relationships to make it happen.

This interlude is a call for established landholders to reach out to new landholders; they may be more receptive to talking and learning about weeds than expected. The research shows that people who are second-home owners are often more attached and feel more at home in their second residence and are keen to connect to the place and its community.[9]

By corollary, this interlude invites new rural residents to consider weeds and other local issues before and once they buy their properties. While often difficult to discover, it is worth investigating the ways in which residents already work together and the possibilities for getting involved. If there are no such groups, then the Local Landcare coordinator[10] can no doubt provide valuable and relevant resources to provide a place to start.

Recipe: Lemon Poppyseed Cake

This is a great recipe that will keep volunteers coming back for more.

Ingredients

Cake

⅓ cup (50 g) poppyseeds
¼ cup (60 mL) milk
185 g chopped butter
1 tbsp finely grated lemon zest
1 cup (220 g) caster sugar
3 eggs
1½ cups (225 g) self-raising flour
½ cup (75 g) plain flour
½ cup (60 g) almond meal
½ cup (125 mL) lemon juice

Lemon syrup

1 cup (220 g) caster sugar
⅔ cup (160 mL) lemon juice (freshly squeezed or store bought)
⅓ cup (80 mL) water

Method

1. Grease deep 23 cm round cake pan, line base and side with baking paper.
2. Combine poppy seeds and milk in small bowl, stand 20 minutes.
3. Preheat oven to 180°C (conventional) or 160°C (fan-forced).
4. Beat butter, zest and sugar in medium bowl with electric mixer until light and fluffy; beat in eggs one at a time until combined. Transfer mixture to large bowl, stir in sifted flours, almond meal, lemon juice and poppyseed mixture.
5. Spread into prepared pan; bake in oven about 55 minutes. Stand cake 5 minutes. Turn cake onto wire rack over tray, pour hot lemon syrup over hot cake.
6. Pour any syrup in tray into jug, pour back over cake.

Lemon syrup

1. Combine ingredients in small saucepan, stir over heat until sugar is dissolved, without boiling.
2. Simmer, uncovered, without stirring, 2 minutes.

I baked this recipe for the 2024 Brogers Creek Landcare Locavore. This recipe is adapted from the orange poppyseed cake recipe (p. 49) in *The Australia Women's Weekly Café Cakes* cookbook published in 2001. Recipe credit: Are Media Pty Limited/Australian Women's Weekly.

Part 2

Eurobodalla

6

Deua Rivercare survives floods and fires

The Deua River might not seem that remote, being only a short drive from the coastal town of Moruya, NSW. But with no access to town water or mains electricity, no convenience shop or post office, and having been cut off time and again by floods and fires, it's not an easy place to live or work.

A drive along the graded dirt road, that winds and wends all the way to Araluen, ~60 km away, adds to the sense of isolation. Houses are scattered intermittently along the road, separated by swathes of Deua National Park and Wandera State Forest, making it difficult to discern where the boundaries are between private property and public lands.

A few kilometres from Moruya, there is a lone Deua River Bush Fire Brigade sign that suggests the presence of a community living in among Deua National Park, but any other signs of their strength and vitality are invisible.

Yet there is something about the 'indescribable beauty'[1] of the Deua River, that connects the residents and inspires them to care for it.

Traditional Knowledge holder accounts suggest that this has long been the case. The Deua River appears in Yuin stories from the Dreamtime, connecting it to the culturally significant Gulaga, the place of ancestral origin for Yuin people.[2,3]

Today, the river provides 60 per cent of the water supply for the Eurobodalla region. It is home to platypus (*Ornithorhynchus anatinus*) and the Australian grayling (*Prototroctes maraena*), an endangered fish species that has been impacted by the 2019–2020 Black Summer bushfires. The catchment also supports 13 threatened bird species and two threatened mammal species that rely on aquatic habitat.[4]

These layers of meaning drive residents to spend their time looking after the river. For over 20 years now, Deua Rivercare has been meeting and seeking government support to remove weeds, making way for the environment to heal from the impacts of mining, farming and gardening.

The Deua River community in the 1970s and 1980s

The relationships and norms that underpin Deua Rivercare almost certainly began over 50 years ago.

For many years the Deua River was a place where 'alternative lifestylers'[5] moved, far from the reaches of Sydney and Canberra. During the 1980s, newspaper articles reported on the latest artist, potter, teacher and timber worker who moved to the area. Each built a mud-brick home[6] or cobbled together houses from recycled materials,[7] some reminiscent of the 'house of sticks and the house of bricks in the story of the three little pigs'.[8] Indeed, residents who moved to the area in the 1980s recount how neighbours would come together to pour concrete slabs and help each other to build their mud-brick homes.

> All the houses up here were built on working bees, and the slabs, the slab parties! And you'd start in the morning, you'd do the work and in the afternoon would be the social – no beers cracked before the [concrete was poured].

What brought the community together in the 1980s was having young families, a desire to be self-sufficient and a need to sustain a low-cost lifestyle.

> Pretty much everybody up and down here was on the dole. So, we were this community that all had limited funds and did a lot

> of things together. So, the community, the whole community spirit was forged back then.

Rivercare members recall the community having its own form of social currency, known as 'Deuas'. Helping one another, through babysitting or other such tasks, would enable residents to earn Deuas that you could then 'cash in' at a later stage.

Newspaper articles note that the community was 'a close, "very social", one'[9] where neighbours would provide one another access to the river through their properties when the river flooded.

Evidence of the social nature of the community can also be seen in accounts of an annual Deua River Bush Race held among the stringy barks at what is now known as Mcgregors Creek. This event was held each year on the June long weekend to raise funds for the Deua River Bush Fire Brigade. Two months of weekly working bees were required to prepare the site for an influx of up to 1000 people. The event drew people to the smell of damper, the sound of bush music and the allure of mystery horse races and a bonfire.[10,11]

Over time, the social activities that bound the community together dissipated. What the interview accounts recall is that as the residents aged, younger generations moved away for work, and new people bought weekend retreats in the area. These changes meant that the old ties that bound the community together weakened; there were fewer reasons to come together.

> They were all young families with young kids, and the kids all used to go to school together and hang out and stuff like that. So, that generation, as in, the adults who first moved there with the young families, or had young families when they were there, they're all a lot older now. They're in their sixties and seventies.

This story of an ageing population is reminiscent of many parts of rural and regional Australia. In the Deua Valley, the median age of valley residents is now 64 years,[12] far beyond the median age of Australians (38 years)[13] and increasing at a faster rate.

The in-between times

While the 'usually isolated' residents often came together to support one another and their local community organisations, there is no mention in the newspapers from this time of the community working together on environmental issues. Interviews with long-term residents confirm that they paid little attention to weeds in the 1970s and 1980s.

Even though many of the residents moved to the area because of their love of the environment, their interest in 'bush protection and biodiversity', 'doing something for the environment' and keeping 'that bushland as intact as possible', they spent little, if any, time weeding. Some admitted to planting exotic plants because they weren't aware of the need for, or benefits of, planting native species. Others admitted to knowing 'nothing about weed control' or that any of their attempts at weed control were largely futile. Some of the newer residents point to the small-leaf privet (*Ligustrum sinense*) hedges around their properties as evidence that the previous owners didn't know, or were unconcerned, about weeds and their impacts.

As Landcare began to gain momentum around Australia in the late 1980s and early 1990s, the residents of the Deua Valley responded to the call for action. In Victoria, where the term 'Landcare' was coined, soil conservation was a key driver of the programme. The same appears to be true of the original Deua Rivercare group, which came together in the late 1990s to work on 'erosion control around the quarry'. Yet there is little documentary or oral records of the group, other than a single mention in a 1998 council report that 'there was a Rivercare project to remove oaks [casuarinas] that were growing in the river ... to stop the banks undercutting'.[14] It was mentioned in passing by current Rivercare members that a group had been established by the Southern Rivers Catchment Management Authority. Lost is any information about how the group formed, who led it, how it operated, who participated or how long it lasted. What the limited information suggests though is that there was latent concern in the community about the river and a desire to do something about it.

The dynamic duo: Alison and Anne

According to legend, there are two women responsible for bringing the current Deua Rivercare group to life in the early 2000s and reinvigorating this remote and ageing rural community. Alison, a new resident, provided the spark, motivating the community to care about weeds and securing the financial resources to support them. Anne, a retired long-term resident, used her time, social networks and sociability to encourage the other residents to join the cause.

Looking back at the early 2000s, before the revamped Rivercare group got going, residents now recognise that a lack of attention and care for the river had allowed weeds to take hold along the riparian zone. Members speak of how they had 'a botanical garden of weeds', how the area 'was absolutely covered with weeds' and that weeds were 'totally out of control', but it was something that they couldn't see at the time. Back then, they were largely plant blind, 'it was all just 'green', we didn't know what [the plants] were!' Those who were more aware of weeds believed that there was little point in trying to control them: 'I think everyone thought that you couldn't get rid of [the privet]. I think no-one had really thought of doing it.' What everyone agrees on is that Alison was instrumental in helping them to see their environment differently and begin to do something about weeds.

When Alison moved to the valley, she quickly became aware of, and concerned about, the extent and density of weeds on her own property, particularly privet. She raised her concerns with one of the locals, who suggested she get in touch with council. The weed officer was quick to recommend the use of herbicides. Although Alison was reluctant, she soon realised that herbicides were the only way to make a dent in widespread weeds as persistent and pernicious as privet. But she also knew that controlling privet would not be a one-person job; she would not only need support from council but also from her neighbours.

Alison began approaching others along the Deua River who had big patches of privet. In doing so, she found others who were passionate about plants. She sought funding to support their initial weed control efforts and educational activities. She soon realised that forming a Rivercare group was the easiest way to enable other organisations to support their work. One of the first organisations to support the new Rivercare group was Eurobodalla Shire Council. The council-funded Landcare Support Officer helped by providing funding and supporting the running of annual workshops and education days.

So the weed work began in earnest and Deua Rivercare got a new lease on life. But who knows how long it would have lasted or what it would have achieved if Anne hadn't come to one of the early field days. From the first event she joined, Anne took it upon herself to oversee the health and safety aspects of the group, taking on responsibility for the sign-in sheet. She soon added the job of recruitment to her list of responsibilities. Being a long-term resident meant that Anne had connections with many in the valley and took joy in visiting people throughout the valley to let them know about Rivercare. As one member explained, they came to be connected to the community and the group thanks to Anne: 'It was through Anne. Until recently I had very little contact with people in the valley ... we're in the middle of a national park.'

In visiting properties and inviting people to join, Anne shaped the norms of the group. She made it clear that any and all contributions were welcome, whether regular or occasional. This inclusive sentiment is evident in a Eurobodalla Landcare newsletter from 2012, where a description of a Deua Rivercare working bee is provided and a call made for others to join:

> New members of Deua Rivercare are always welcome. Participation at one work morning does not mean you are expected at every work morning. You will be more than welcome whenever you are willing and available.[15]

There is one other thing that Anne introduced to the group that was essential to its functioning and longevity: morning tea. With her long-term connection to the community and familiarity with the activities that had previously helped the community to thrive, she realised the value of encouraging social connectedness. Working bees needed to be about more than controlling weeds. And so residents were encouraged to come along to the working bees, even if it was just to join for a cup of tea and some cake.

There isn't a single interview I conducted where Anne and Alison's dedication went unmentioned. In each conversation, it is clear how these two women have inspired the community and those who work with them to care for one another and the Deua River.

> [Alison and Anne are] just amazing. They're real role models, not just for the community but for us as well. These guys are doing this stuff off their own bat ... They're not getting paid for it. They're just doing it because they love doing it and they want to make the place better. That's pretty amazing.

Emma

There is at least one more woman whose work in supporting Deua Rivercare must be acknowledged. Emma has been the Eurobodalla Landcare Coordinator for almost a decade and, unlike other Local Landcare Coordinators across south-east NSW, and NSW more broadly, has the benefit of being hosted by the local council. While this can come with challenges – many rural residents have negative opinions of their local council – it also means that Emma is able to identify and mobilise council resources to support the group. Such support means that Deua Rivercare has been able to partner with the local council in its grant applications and count on their support in prioritising weedy work, approaching residents, running workshops, organising access to resources, managing weeds on council land and promoting the work of the groups. In these ways, Emma supports Alison and Anne with the tasks of accessing financial resources and managing group dynamics.

Emma also helps Deua Rivercare to liaise with and gain the support of other government agencies, such as NSW National Parks and Wildlife Service, Forestry Corporation of NSW and the South East Local Land Services. This bridging role that Emma plays is greatly valued by members of Deua Rivercare. Residents often find NSW National Parks and Wildlife Service personnel difficult to engage with because of a high turnover of staff and because of the limited resources they have available to manage the size of the national parks and nature reserves that they are responsible for in the valley. Emma, as staff of the local council, is able to make time for meetings with new staff and explain why certain weeds are a priority for residents, as well as the opportunities for supporting one another's work.

Beyond Deua Rivercare, Emma is responsible for supporting the 20+ Landcare groups that form the Eurobodalla Landcare network. In this role, she facilitates meetings among the groups so that they can share experiences and learn from one another. When possible, Emma joins in working bees, lending an extra set of hands, a wealth of knowledge about how best to control weeds, especially the latest arrivals, and helps with monitoring progress. Emma also helps the groups with data collection, synthesis and report writing that needs to be done once grant funding has been secured. This takes a huge burden off local groups with respect to the administrative tasks and frees them up to spend more time doing the things they love.

A framework for collective action

Many aspects of Deua Rivercare's work, like that of the other Landcare groups, can bring to life academic concepts about how community groups collaborate to manage environmental resources. Some of these key ideas include the need for a shared goal, clearly defined boundaries, addressing unequal contributions by group members, how to share communal resources, navigating tensions and feeling a sense of progress.

When groups come together to work collaboratively, there are many ways in which they can support one another. First and foremost, there needs to be a common goal.[16] Deua Rivercare was quick to recognise that what the members all had in common was a desire to improve and protect the riparian zone of the Deua River. This meant that they would focus their effort on the 'habitat-changing weeds': weeds that have a transformative effect on the riparian corridor by crowding out native species and impacting water quality. This includes woody weeds such as privet, wild tobacco (*Solanum mauritianum*) and cassia (*Senna pendula* var.

glabrata), and vines, including cape ivy (*Delairea odorata*), moth vine (*Araujia sericifera*) and Madeira vine (*Anredera cordifolia*). Some widespread riparian weeds, like wandering trad (*Tradescantia fluminensis*) that grows prolifically along the riverbanks, were recognised as a problem but for which weed control would only exacerbate their impact because spraying it 'would just denude the whole joint'. So the group aimed to maintain the river ecosystem by enhancing riparian vegetation.

As a next step, it is important for groups to clearly identify the boundaries of where they will work.[17,18] In the Deua Valley, the focus is on the areas close to the river, with the valley sides providing a clear demarcation of the limits of where the group would work. Initially the pressing questions were how far up-river to work and where to start. The answer to the question of how far to go evolved over time. Initially, the group started in the lower reaches of the river, then the upper and finally the middle. Bit by bit they pieced together funding to carry out weed control along a 42 km stretch of river, from Araleun Creek to Burra Creek.

Related to setting a goal and determining the reach of a group, both in terms of area and potential members, is development of expectations of group members. It is important to recognise that not everyone is equally passionate or able to contribute to the common goal. In academic research, this is known as the 'weakest link' challenge. Finding a balance can be tricky. In many groups I've worked with over the years, members harbour resentment towards those who do not actively contribute to the shared goal through labour or regularly attending meetings but still receive support, especially financial support.

It is in this regard that Deua Rivercare seems to have created an environment where funding is available to support anyone along the 42 km stretch of river, regardless of their participation in the group. This appears to be because many members have property where the weeds are hard to reach and because the group recognises that achieving their goal would not be possible with working bees alone. Working bees are clearly important for building a community of care for the river, but there are many areas that are hard to access or where the weeds are so thick that specialised equipment and contractors are needed.

But contractors don't come cheap. It is here that Alison's ability to secure funds has been critical. Grants from NSW Fisheries and the NSW Environmental Trust have been instrumental in paying contractors to help Deua Rivercare with its weedy work. The contractors are referred to as the 'Hard Core Team' because of their enthusiasm, their ability to work in hard-to-reach places and the immediate benefits that can be seen with the work that they do: 'They got so much work done. I've never seen a team of contractors like that, and they just saw it as a challenge.' The contractors help to keep motivation going, and the group inspire the contractors with their passion for and knowledge of the local environment.

The contractors who work in the Deua Valley, like many of the contractors who work for the Landcare groups in this book, add value because they've been working in the area for a long time. Being involved with the Rivercare group for almost a decade means that they know the terrain and the weeds, when is the best time to access different parts of the river and when is the best time to control the weeds. The contractors also have relationships with many, if not all, of the landholders, which makes getting permission to work on the properties easier.

This speaks to the other part of the challenge of getting everyone to contribute to collective action dilemmas. Having funding is not necessarily enough to convince those who are not involved to allow Rivercare or contractors onto their property. It also requires an ability to navigate neighbourly tensions. Non-members may not wish to engage with Rivercare because of an aversion to the use of herbicides, a desire to be left alone or concerns about letting unknown people onto the property.

Two of the people I interviewed indicated that they enjoy being by themselves and don't consider themselves to be social people. They don't ordinarily join volunteer groups, but they explained that Deua Rivercare provides a relaxed space where they feel comfortable and it also gives them an opportunity do to something to help other people, a 'purposeful endeavour'.

These stories suggest that Deua Rivercare has managed to achieve what academics call 'friendly distance',[19] or what members term 'country reserve' and 'neighbourly assistance'.

Sociologists have noticed a particular tension that exists among neighbours; a 'good' neighbour is one who 'respects other's right to privacy but who at the same time makes herself or himself available to be called upon if necessary'.[20] Often neighbours find it difficult to strike a balance between being sociable but not intrusive. In Deua Rivercare, members recognise and respect that people have moved to the area for peace and quiet, that tensions are inevitable, but they are also ready to lend a hand at a moment's notice.

> We don't live in one another's pockets. We may not see one another for a week or so sometimes, depending on what we're all doing, but if the word gets out that someone needs a hand, it just sort of happens, and so therefore if someone's got a weed issue, that will be sort of mentioned, and it's a case of, 'Well, you're the next work day.'

Beyond the challenge of maintaining neighbourly relations, another reason why residents may not join Rivercare is because they feel that weeding is futile.

> I was talking to another neighbour ... she's very, opposed to Rivercare sort of operations because she just feels as though it's wasted effort. That you remove the weeds, and no sooner have you removed them than they're back again, so she just doesn't quite get why people would persist with it.

There are two ways Deua Rivercare addresses such concerns. First, as indicated above, they not only work on weed control but on creating a range of shared benefits, such as learning about weeds and the natural environment through information sessions, as well as creating a 'connected community', which are valued by many and are certainly a drawcard for some new residents.

> The first thing I did, was to link up with the Deua Rivercare as a way of creating a community for myself as well as wanting to learn more about how we should be caring for our property and

> our land ... the benefits are the people that you meet along the river, and the connections you make with the community, and there are some amazing people who live along the Deua.

Beyond providing shared benefits, the group is proactive in recording and promoting their achievements over time in reducing the weed burden and enhancing the riparian zone.

Kayak monitoring

Deua Rivercare has an innovative way of establishing how weedy the riparian zone is and how it is changing over time: they do it by kayak.

It all began one year when the river was low. A family member was visiting and had a bit of extra time on their hands. They used that time to walk the length of the river and photograph and map the weeds along the way. It was through this exercise that the group discovered patches of Madeira vine that they were previously unaware of and obtained a better sense of where all the patches of weeds were.

After that, the group secured funding and the support of council to do the same monitoring by kayak. Going by kayak meant that they didn't need to navigate property boundaries and locked gates. In four days they could cover their whole area and spot the weeds that are often hard to access. The first time they kayaked the length of the river in 2012 it was a low-tech enterprise. They had topographical maps and noted on them where the weeds were, supported by photos and vegetation transects. This trip showed that 'there was almost one long and continuous stretch of invasive species along the river'.[21] The maps provided essential information for prioritising where and what kind of effort to invest in weed control.

Over time, the monitoring has become more sophisticated. The council provide the GIS technology and staff to facilitate the mapping every second year. The group now have the kayak monitoring down pat. With two people in each kayak, one is responsible for spotting and identifying weeds and their density and the other is responsible for recording the location – patch to stretch – of weeds on the GIS.

There is a joy that is evident when members speak about the kayaking trips. Members can see for themselves how the vegetation is changing and the difference that they're making.

> I remember the first year that we did it, we would get to certain sections and just go, 'Oh my God, how? This is ridiculous!' We'd start the point there, and we'd kayak a kilometre or so before we could end that point ... just the banks are covered with sycamore, the banks are covered with blackberry. And various other bits and pieces. And now a lot of those properties you can go back to, and you can spot [spray] them and just have backpack on. So yeah, that's pretty special.

The kayak monitoring means that the group can confidently demonstrate that over a six-year period, ending in 2018, the group's almost 5,000 volunteer hours and associated contractor time achieved a 90 per cent reduction in woody weeds[22] and an improvement in overall catchment health.[23] 'Things like privet and sycamore, tree of heaven, they're all our original targets. There's hardly any of those left. It'd be very isolated occurrences.'

On fires and floods

Like all the groups in this book, the Deua Valley didn't escape the Black Summer bushfires unscathed. In late December 2019, the fires came to the valley. Come 4 January, there was no stopping the Clyde Mountain fire 'roar[ing] over the ridge'[24] and sweeping through the valley, destroying six homes along the way and burning along the riverbed. After this terrible day the fire continued to burn less intensely until the wild winds of 23 January brought the beast back to life and another four homes were lost.

Most residents fled to Moruya for the month of January, some staying with members of the Rivercare group who live in town, others in evacuation centres or vans. While many members of Deua Rivercare are also part of the Deua River Bush Fire Brigade, the fire storm that roared through the valley prevented them from being able to defend their own properties. The next day they pushed through the fire front to evaluate the damage. Because Rivercare members had spent so much time working on weeds on one another's properties they knew the terrain well and the location of key infrastructure, so those who were members of the Rural Fire Service benefited from having intimate knowledge of the properties along the river. Their close social networks meant that they were able to rapidly relay messages back to those in town about whether homes had survived.

While destruction of homes, cars and other infrastructure was a big loss, residents also had to contend with the shock of the burnt-out, 'apocalyptic' landscape. The mountains all around the valley were burnt and the trees were 'just dead sticks on the slopes'. The sense of loss at the scale of environmental change was still acute two years on.

The ability for the group to support one another through the aftermath of the fires was hampered by the COVID-19 lockdowns in early 2020 and then by multiple floods. By the time I visited the group in March 2022, they had had 10 floods since the fires. The flooding not only damaged the riparian area, it also made it difficult for some residents to access their homes, and it split the valley for 16 months where a landslip collapsed over the road in two places. Even once the lockdowns had finished, many working bees were called off because the area that they had planned to work on was under water or because it was impossible to cross the river.

As soon as they were able, the group decided to swap their usual working bee with a post-fire vegetation management session to help a neighbour who had lost their home in the fire. Members brought their tractors, slashers, chainsaws and whipper snippers to clear up the property and make it manageable.

> [The Rivercare meeting was held where] they lost two houses up there in the bushfires, and there was a really big turnout of locals ... and we went up there and planted a lot of trees and shrubs.

The Rivercare group also joined forces with a local scrap metal business to coordinate the removal of burnt-out cars and other detritus that weren't covered by the government-funded clean-up programme.

> We had a huge clean-up along the river, so that was really cool ... people feel really supported with things like that. It builds the goodwill for the Rivercare.

Beyond these events tailored to bushfire recovery, members make it clear that even the regular Rivercare working bees have been crucial in providing emotional support to one another through this challenging time.

> [While weeding] you have a conversation. You catch up with someone that you haven't seen for a month or two months or three months, and you're finding out about their family. We all know about one another's families. We all know about our kids and grandkids. We all know who's been sick, who's died. All those things that – it's not just the pulling of the weed or the poisoning, the dabbing. It's the conversation and the company.
>
> We don't make a point of saying, 'Oh, we'll go and have a cup of coffee.' We say, 'Let's go weeding.' So, we weed, and then we have the cup of coffee afterwards, or whatever, but by that stage, you've caught up with that particular person, and you know sometimes people will talk about what's worrying them, and it's confidential a lot of the time. It's something that they want to share but they haven't had the opportunity to sit with someone. And sometimes there's tears about something halfway through the bush, and they'll tell you something that's been worrying them, or someone's really ill, or something. It's a comforting time as well.

This quotation shows the importance of the social connection that Anne fostered. The morning teas and side-by-side weeding have provided an important dimension to the social fabric of the community, one indispensable to the other.

With respect to the weeds, in some ways Deua Rivercare members were more positive than other groups about the impact of fires on some weeds. They noted that since the fires, there had been much less cape ivy, something that they had previously considered to be one of the 'worst weed[s]' and that they had spent hours removing, had largely disappeared since the fires: 'we're virtually cape ivy free'. Contractors and council staff made similar observations. There was a sense of relief that they no longer had to contend with these transformative weedy vines.

There were other weeds though that thrived in the post-fire environment, especially wild tobacco and turkey rhubarb (*Rumex sagittatus*), and others that thrived post-floods, like wandering trad. The turkey rhubarb has been challenging because residents and contractors are less familiar with it and have needed to experiment with new ways of controlling it. This process of experimentation has been slowed by the cumulative and compounding natural events that have prevented the group from meeting.

With excessive rainfall, COVID-19 lockdowns and floods preventing the Rivercare group from meeting, these plants rapidly spread throughout the valley. Not only were Rivercare members unable to spend time weeding after the fires but also contractors and council staff. In part this was due to the inaccessibility of sites, but also because contractors had more work than they could possibly manage. While funds were available for post-fire weed control, there weren't enough people available to do the work. This is particularly frustrating for locals because the fires provided a unique opportunity to get control of some weeds before they became re-established.

> I know after the fires, if we'd been able to get in, if we had access we would have been able to have a lot of things under control. So, there's frustration that there was the landslides and the floods and all the things that stopped us getting there … I have driven up there and felt really overwhelmed and really frustrated, and kind of even angry at myself. Like damn, why weren't there funds when we needed it? Why couldn't we get up here? There's a point where we didn't have contractors available, because there was that much remediation after the fires.

Burnout

When I first met the group in 2022, it was at an emergency meeting. No fires were blazing, the river wasn't in flood, the road cut by the landslide had finally been remade and we were now 'living with COVID-19'.[25] But the community was reeling. Their passion – for working bees, for caring for the river, for coming together – had been reduced to embers. For a while now, working bee numbers had been 'underwhelming', with sometimes only one person turning up. If the group was to survive, something would need to change. But what? And who? And how?

It was time for some introspection. So on the first Saturday of this particular month, instead of hauling the tool trailer to a work site, the group gathered at one of the member's homes with cups of coffee and home-made cakes. It was the best turnout in months, and many more had emailed to say that they were committed to seeing the group continue.

Something needed to change. The leadership needed support and the way of working together needed to adapt to make sure it continued to suit as many people as possible. For two hours the members discussed the challenges the group faced: the lack of younger residents and members; the decline in volunteer numbers, not only in the Rivercare group, but across a wide range of community organisations that members belonged to; the challenges of reaching weedy sites; and the barrage of emails that members receive from many sources, making email communication less effective than it once was.

They also talked about successes, the working bees that had the most participation, the other events, like weed identification days that were valued by the group. And the overall impact that the group had had, from the early days when few people were concerned about weeds, to the evolution of the group, the grant successes, especially the longer-term funding that enabled the group to finally approach all the landholders in the valley, and the awards received for their work.

Newer members relished hearing these stories, knowing more about the group they had joined, about their passion for the environment, and that they weren't alone in their learning journey, that the people they now knew to be highly knowledgeable about weeds were once just like them in not being able to tell one plant from another. They learnt about the contractors, about the work they did in supporting the group and how they documented the health of the river through kayaking trips. They also heard about other community groups who came to help over time, high school students and Aboriginal groups who came to do plantings.

The wide-ranging conversation inspired fresh ideas, new ways of working as well as re-incorporating past successes. Members volunteered to help organise working bees and agreed to a new way of committing to one another. People would now let the person organising a working bee know if they were planning on coming, so events could be cancelled ahead of time if numbers were too low. There would also be more diverse events, not just working bees, but plant identification walks with local experts, days to talk about good chainsaw practices, seed collection days and social events with other community groups, such as a movie night at the fire shed and of course the annual Christmas party.

The first issue of the *Deua Daze* produced by one of the members later that same year is testament to the success of this emergency meeting in reinvigorating the group. Not only was it the first time the group had published a newsletter in years but the newsletter also reported on multiple well-attended events that had taken place in the months following the meeting. A walk with botanist Jackie Miles had been over-subscribed, a dozen had met for a working bee at a local farm and the mid-winter family movie night at the fire shed was well attended. Deua Rivercare was getting back on track.

Weedy working bees

I rolled up a bit before 9:30 a.m. The telltale sign of cars parked between the dirt road and the fence confirmed I had the right place. Two hopeful horses were at the fence to welcome me.

Just beyond was the distinctive trailer that proudly announced Deua Rivercare was here to work. The side flaps open to enable group members to get their equipment, gloves and chemicals easily, if desired. Group members are also often spotted with their long-sleeved forest-green Eurobodalla Landcare polo shirts sporting the motto 'I'm part of the solution'.

While members arrived and signed in, we got onto the important task of placing our cakes and other tasty treats on the table and applying our choice of insect repellent to keep the ticks at bay. Having had two ticks find me most tasty in the last year, the first of my life, I eagerly and liberally applied repellent.

Once everyone had arrived, there were introductions and an explanation of the task at hand today – the pesky wild tobacco that was thriving after the fires – and the need to be wary of wombat holes and snakes. Gaiters were distributed to those who wanted them. A story of some recently encountered red-bellied black snakes (*Pseudechis porphyriacus*) made the gaiters a popular choice.

Those who were more experienced provided a demonstration of how best to control wild tobacco, pointing to the importance of cutting the plant as close to the ground as possible and applying herbicide within seconds of cutting it. They also talked about cutting and bagging the seed pods, to prevent further spread.

We then split into twos and threes to make our way along the riverbanks. As we began to work, we settled into a pattern. One of us would cut the wild tobacco, another would apply the herbicide, and if there was a third, they would collect the seed pods. And then we would find the next wild tobacco plant popping its head above the other vegetation. Often, we would have to push our way through shrubs, go over logs and through ditches to get to the base of the plant and take it down in one fell swoop.

We would stop along the way as members pointed out the white cedars and mulberry trees. They would ask one another what unfamiliar plants were, with some brave enough to taste them as a form of identification!

Once we'd spent a good hour and a half along the river, with a dozen or so bags laden with wild tobacco seed pods, there was collective agreement that it was time to start making our way back to the morning tea site. We could already see the impact our morning's work was having as we walked back to the paddock. The culled wild tobacco trees were wilting, already beginning their process of decomposition, their potential to spread halted for now.

But our weeding work wasn't quite done yet. Never ones to walk past weeds, as we made our way through the paddock, we worked in an emu line to pluck fireweed (*Senecio madagascariensis*) diligently and methodically, bagging the weed as we went.

Some went on ahead to boil the kettle and make sure we were all rewarded with a cuppa and a spread of cheeses, crackers, fruit, nuts, dips and cake. The carrot and ginger cake was scrumptious.

As we sipped and munched, the conversation turned to the property and its setbacks of late, highlighting the value of the work that had been done.

Members contributed titbits of information about the health issues of members who weren't present, the bush telegraph working its magic. Other environmental initiatives were also discussed, tree plantings and plant identification workshops that members could attend.

And with that, it was time to pack up and go home, invigorated by the work achieved and ready for the next event in a month's time.

A new way of working

Since the emergency meeting, the group seems to be finding its way again. The weight of responsibility for leading the group no longer rests on Alison and Anne's shoulders. Others have stepped up to organise diverse Rivercare events. There is also a new way of communicating, with working bees called off when the weather isn't suitable, numbers are too low or alternative are plans made to suit those who are available.

So far, this more flexible way of meeting seems to be working. One of the working bees I attended in 2024 had over 20 members in attendance.

The 'occasional newsletter of Deua Rivercare' is also testament to the dozen or so members who have attended working bees where they have learnt how to use the iNaturalist app for plant identification or showed up on Clean Up Australia Day to remove all manner of odd waste, from soiled nappies to vape packaging to lone shoes.

The challenges the group faces are not unique. Getting new residents to join, enticing younger members and sharing the load are a common challenge to almost all the groups in this book. Deua Rivercare, like Towamba Valley Landcare (Chapter 9), found that calling an emergency meeting was an effective way to reconnect with their purpose, speak to the meaning of the group and find a new way forward.

This process of fading and reinvigorating feels inevitable and is part of the longer story, not only of Deua Rivercare but of the sense of community in the valley.

This chapter began by talking about the rise and decline of community connectedness that came and went with the slab parties. Yet there are earlier stories of teamwork, dedication and care that speak to the ways in which people have come together over the course of centuries, if not millennia.

Aboriginal Knowledge Holder accounts of the Deua River speak to the way women have worked together and cared for the river. A book about the Deua tells the following story about the spiritual significance of a women's site (at what is now known as Bakers Flat):

> The story was explained to me, if you go to Bakers Flat you can see that the rocks lined up, they look like women kneeling into the water. I was told they were washing the roots of the bull rush and setting the dilly bags to clean the burrowang seeds. There was suppose to be one woman on the side of the bank keeping guard, making sure no one would sneak in. On one occasion the woman lay resting in the sun and fell asleep. Some men came past and noticed that the woman keeping guard was not doing her job, so the men turned her into stone as punishment for not being vigilant. Someone in my family told me that story when we were camping at McGregor's Creek... .[26]

While the women in this story were prevented from continuing their caring work, the story of their efforts lives on, inspiring others to learn from their experiences.

There are also colonial stories of women helping women to survive in this remote valley. In one account titled 'Courageous women of the

Deua River'[27] it is evident how two women inspired others through their tenacity, resourcefulness and skill in establishing their cattle farm miles away from the nearest town of Araluen. These two women were not only admired for the way they single-handedly built their own farm but also for the way in which they supported other women who faced calamities.

While the stories of these Aboriginal and colonial women are rarely, if ever, revisited by the local community, their presence speaks to the inevitable rise and decline of community connectedness, to the potential for it to be re-established when there are those willing to take the initiative, and the long-term respect and esteem that will follow such care and compassion. The question that has only recently been answered is: who will step forward next?

INTERLUDE: BUSHFIRES AND WEEDS GO HAND IN HAND

The massive fires that spread across Australia from September 2019 to February 2020, known as the Black Summer bushfires, burnt more than 11 million hectares. The scale of loss was unprecedented. Billions of animals were killed, thousands of homes were destroyed and more than 30 people lost their lives.[1] The images made international headlines with apocalyptic scenes of coastal residents fleeing the mainland in boats or being rescued by the Navy.

Yet the role that plants – both natives and invasive – played in the bushfires and the impact the bushfires had on plants and people's relationships to them is noteworthy for the silence in public fora about this.

In a media analysis of the Black Summer bushfires conducted by researchers at Monash University, three key themes were apparent in the news: the unstoppable power of nature; the triumph of humanity; and the loss of biodiversity.[2] In their 30-page report, analysing the content of 700 articles (only 1 per cent of all articles published in the 5-month bushfire period), the term 'plants' only appeared once in relation to loss of species. Parched 'vegetation' was also only mentioned once.

Weeds give way to bushfires

'Parched' goes only some way to conveying what had happened to vegetation in the months and years before the bushfires. Preceding the bushfires, there had been widespread and unprecedented drought. Countless crops failed due to a lack of rain to the point where some farmers didn't even bother sowing crops.

The conditions were ripe for some weeds to spread. Weed seeds were carried hundreds of kilometres in dust storms and hitched rides in fodder that was transported interstate. And some weeds happily grew where crops and pastures refused.

> The situation is now described as a 'green drought', where paddocks look green from a distance, however pasture growth

> is hamstrung by low rainfall and grass growth is stymied by weeds... A lot of what's being grown in paddocks at the moment is weeds and they have little to no nutritional value for stock.[3]

This was the backdrop to the Black Summer bushfires: weeds spreading and thriving, driving vegetation change.

This context is important because we know that climate change will make droughts and bushfires more common. We know that some weeds have a competitive advantage over native plants,[4] and that their competitiveness is likely to offer them advantages when faced with climate change.[5] Together, this means that understanding plants, their condition and spread needs to be central to our conversations about bushfires.

Bushfires give way to weeds

In February 2020, a group of 28 weed researchers, policy-makers and practitioners met at Charlotte Pass in Kosciuszko National Park to discuss the future of weed management in Australia.[6] The Black Summer bushfires were still raging and the group was worried about the many ways weeds were likely to be spreading into new areas.

A media release written by the group that same month explained: 'Weeds may have come in on some of the emergency vehicles and machinery during the fire control response, or may have pre-existed but are now exposed by the fires.' The high winds, like the dust storms during the drought, were also likely to have carried in weed seeds from afar.

The group of experts called for action in the months immediately after bushfires 'to detect and respond to new weeds or fire-promoted weeds to stop them before they affect highly stressed and threatened species'.[8] But their concerns largely fell on deaf ears, with significant consequences.

The emotional toll of bushfires and weeds

While some funding was made available for weed control work as part of bushfire recovery efforts,[9] the amount of funding and support provided was insufficient.

The stories collected for this book show that bushfire-affected landholders throughout south-east Australia experienced grief, a sense of being overwhelmed by weeds and fear due to their experiences with plants during and after the bushfires. The lack of government support and information available on how to deal with weeds post fire exacerbated this emotional toll.

Grief

The loss of vegetation, alongside the loss of infrastructure, was acute.

> Nothing was green. The house, the garden, the roundabout. It wasn't very pretty. Swimming pool. The old cottage. It was devastated. The old shed went. I mean, that was just black. All of that was black. It looks fairly black now, but it was way blacker.

The sense of grief was about more than the loss of plants. There was a sense of loss for what had been a familiar landscape, the work that had been done in weeding and planting, the hours of work done.

To really understand how much affection people have for trees they have planted, I share here a quotation from my PhD research. I will never forget one landholder I interviewed who had lovingly planted many trees on her properties. She referred to those trees as her babies.

> There wasn't any of those trees [when I moved here] ... I had to walk up and plant everyone. I had to dig a hole, cart the water up in buckets. They're all my babies, every new tree is my baby.[10]

Thankfully her property was not affected by the Black Summer bushfires, but her affection for those trees has stayed with me. It reminds me of the significance of plants, how much they mean to people and the deep grief they must have experienced when those trees went up in flames.

Bereavement support for bushfire affected landholders focuses on loss of property, infrastructure, lifestyle and routines. There is little, if any, acknowledgment of the grief associated with loss of landscape.

In future, bushfire support needs to acknowledge explicitly the grief that comes from our relationships with plants.

> Everyone was devastated after the fires, particularly with the big forest casuarinas. Those massive things that sit on the river there. Quite a few of them died, and people were so upset. I was upset as well. There was a massive river gum that was on the corner of a bend there, and that was burnt out and fell over. People were just so upset about that.

Overwhelm

While still grieving the loss of pre-bushfire vegetation, landholders began to feel overwhelmed because of the weeds. With the three wet years that followed the bushfires, weeds made the most of their competitive advantage. Many interviewees commented on the emotional burden that they or their neighbours experienced.

> A friend has a property that completely burned ... And I've heard a few people say this, 'I can't, I'm just overwhelmed with the weeds.' Because everything went bare and then [there were] large amounts of rain and then the weeds [came]. And the weeds are taking up acres, acres ... Tall fleabane, higher than your head ... They're overwhelmed with it. And they want to plant back gardens and get your paddocks back that aren't just weeds.

In part, the sense of being overwhelmed came from the number of weeds that were thriving and their volume. It was an impossible task to keep on top of weeds while also clearing the burnt debris and making plans to rebuild. There was also the mental load associated with being vigilant for new plants and the effort required to find out whether they were a weed and likely to become a problem.

Feeling overwhelmed came from landholders realising just how out of their depths they were regarding their knowledge of plants and how plants respond to bushfires and floods. Were these new plants going to be short-lived phenomena or would they become their weed burden for years to come?

Bushfire recovery staff were ill-equipped to answer questions about weeds and local council weed officers couldn't keep pace with the deluge of inquiries. The bushfires highlighted the fact that there aren't enough people with the botanical knowledge required to support landholders with weed identification. While there are many apps available for this task, landholders don't have enough confidence in the identifications they provide. For example, some of the apps are designed in the United States of America and so the plant information is not always accurate or relevant.

Even where council staff had created videos[11] about weeds of concern after the fires, they discovered that landholders wanted a more interactive experience. Bushfire-affected landholders wanted to have a conversation with someone knowledgeable about plants and weeds. They wanted someone to be able to see their property and have a discussion with them about weeds and where to prioritise their efforts.

It was suggested that in future it would be beneficial if bushfire recovery included workshops for landholders,

> where you have a really good botanist to go out with a bunch of landholders, maybe in every catchment, and just do a walk-around. Show them what's a weed, what's not, what's worth worrying about.

Fear

In amongst the grief about the loss of plants and feeling overwhelmed by the volume of weeds that needed to be controlled was fear that the new plant life would feed the next bushfires. This fear fuelled a rethink about what constitutes a weed.

One of the plants that received much attention up and down the coast was black wattle (*Acacia mearnsii*), a native plant. Residents observed that places where black wattle was dense burned especially hot, leaving the earth scorched for months afterwards. These patches refused to turn green. While most landholders were aware that black wattle is a pioneer species that will senesce in due course, they were worried that the next lot of bushfires will come through before then and will be exacerbated by the black wattle. Black wattle was now an unwanted plant.

> It's a real concern that in 5 years' time we're going to have a landscape in southern New South Wales that's covered with thousands of hectares of black wattle senescing and, you know, basically a fuel bomb.

Again, the existing resources available to landholders were insufficient. Any resources about weeds after fires didn't take discuss native plants; no noxious weed list in Australia includes native species. So who can landholders turn to in the absence of such information?

The Landcare lifeline

For all three emotional responses – grief, feeling overwhelmed and fear, Landcare groups have provided an important service where there was limited government support or information available.

Landcare groups supported one another through the grief of losing beloved vegetation. Working bees in the years after the fires not only helped to control new weeds but were also central to clean-up efforts. Landcare groups provided a forum where landholders could seek advice about plants, get help with weed identification and come up with strategies for dealing with new weeds, regardless of whether they were native or invasive.

Yet it is important to note that people who were not members of Landcare before the fires found it difficult to access weed identification information and weed control advice through Landcare after the fires.

> I've rung Landcare through Bega Valley Shire Council four times. Oh, we'll put you in touch with this guy up at Verona, he knows your area. No one's rung me back.

This suggests that while Landcare provides important post-bushfire support on plant-related matters, the extent of the support may be limited to existing members.

Preparing for the next bushfires

As the recovery process continues, I wonder whether we will foreground the role of weeds and their emotional toll on landholders when the next lot of bushfires inevitably come through.

Will we provide training to the Bushfire Recovery staff on what plants landholders should be concerned about and ones that they can just let grow? Will we arm BlazeAid[12] volunteers with training in weed control, so that once they have reinstalled fences they can support landholders with the mammoth task of weed control?

I wonder how we might be able to use the experience of the Black Summer bushfires to highlight the role that weeds played in the fires' severity and show just how significant this often-overlooked problem is.

Recipe: Banana Cake with Passionfruit Icing

This cake is good for emotional eating. I recommend accompanying it with a cup of tea or coffee.

Ingredients

Cake

125 g butter
¾ cup (150 g) firmly packed brown sugar
1 tsp mixed spice
2 eggs
1½ cups (225 g) self-raising flour
½ tsp bicarbonate of soda
1 cup mashed banana
½ cup Greek yoghurt
¼ cup (60 mL) milk

Passionfruit icing

1½ cups (240 g) icing sugar
1 tsp soft butter
2 tbsp passionfruit pulp, approximately

Method

1. Preheat oven to 180°C (conventional) or 160°C (fan-forced).
2. Grease 15 cm × 25 cm loaf pan, line base with baking paper.
3. Beat butter, brown sugar and spice in medium bowl with electric mixer until light and fluffy.
4. Beat in eggs one at a time until combined.
5. Transfer mixture to large bowl, stir in sifted flour and soda with remaining ingredients.
6. Spread mixture into prepared pan.
7. Bake in oven about 50 minutes.
8. Stand cake in pan 5 minutes.
9. Turn onto wire rack to cool.
10. Spread cooled cake with passionfruit icing.

Passionfruit icing

1. Place icing sugar in small heatproof bowl.
2. Stir in butter and enough pulp to give a firm paste.
3. Stir over hot water until icing is spreadable, do not overheat.

I made a double batch of this cake to take to a Deua Rivercare working bee and an Upper Kangaroo Valley working bee that were held two days apart. I usually use a round cake tin and make 1.5 times the amount of icing for a single cake. This recipe is adapted from *The Australia Women's Weekly Café Cakes* cookbook published in 2001, p. 20. Recipe credit: Are Media Pty Limited/Australian Women's Weekly.

Part 3

Bega Valley

7

Better without bitou bush: a story of Tathra Land Care

If you happen to be in Tathra (~20 km east of Bega) on the fourth Saturday of the month, somewhere nearby will be a group of passionate volunteers scouring the dunes and hinterland for bitou bush (*Chrysanthemoides monilifera* ssp. *rotundata*), a woody weed with yellow flowers. If you're close enough to where they're working, you may spot them clambering over fallen trees or under dense foliage. But most of the time they are near impossible to spot among the trees, shrubs and vines. The only evidence that they are nearby is the occasional voice calling out: 'Are you OK there?' Or, even rarer still, a gleeful shout: 'I've found one!' if they happen to come across one of the last remaining bitou bush plants in Tathra – the absence of which is the ultimate sign of the group's success.

Bushwalking with a purpose

At 8 a.m. each month, cars slowly arrive at the designated meeting point. Jim is always the first. The boot of his car is ready to go with gloves, mattocks and a sign-in sheet. Occasionally an A1-sized map of the area is used to show the group where they will focus their efforts that day.

One by one, members gather around, catching up on the latest happenings around Tathra. Before long, a dozen or so keen volunteers are itching to get going. Once everyone has arrived, it's time to head off.

For 30 years, Tathra Land Care has been meeting on the last Saturday morning of every month from March to November as long as it's not raining. There have consistently been between 8 and 12 people at each two- to three-hour working bee, a number that's described as 'not too many and not too little'. There is a familiar pattern to where the group meets and how the members pair up to go looking for bitou bush – a plant that crowds out native vegetation and accelerates erosion.[1]

Each year the group starts by working from south to north along the foreshore, taking three months to cover the length of the beach. They then work through the hinterland in the remaining months, so that each area is covered at least once every two years. These monthly working bees have been described by multiple members as 'a bushwalk with a purpose', although having been to a few, it's less of a bushwalk and more of a bush obstacle course. Dune vegetation may look like it's easy to navigate but you soon learn that it is hard to find your way through the maze of turning vegetation. My 40-something-year-old self found it hard to keep up with Jim, who must be close to twice my age.

The revegetating bushland at the top of the hill behind Tathra is no walk in the park either. Aside from dense stands of wattle that have grown back since the 2018 bushfires, there are deep gullies that make the 'walk' slow going. At times you can't see where you're stepping and just have to propel yourself forward, hoping that your feet will keep you upright. As for the boggy swamp between the dunes, in the swale, you'll almost certainly get wet feet. Oh, and watch out for the wombat holes!

In joining the group, clambering over fallen trees and ducking and pushing your way through thickets of wattles, it becomes apparent why its members enjoy this 'walking off-track in the local environment'. Over time you come to know the plants and the contours, the sounds of the birds and the ocean. One member described how 'I got to see different vegetation communities, different geologies, different landscapes, creek lines, rock formations'.

There are also native animals that they come to know:

> I always say, 'Oh, I'm expecting a lyrebird now, that should be in here!' And what do you know, there it is. Just that sense of anticipation as you move through the landscape, and the things you see.

In a video from 2001, a group member talks excitedly about a 2 m diamond python (*Morelia spilota spilota*) that they saw 'just sitting on the rock with three or four little ones'.

There is also an opportunity to learn from other group members about the local environment and how it changes.

> [W]e always found that if you say anything, 'right, you've got Bega dry grass forest here, and here are indicator species to tell you you're in this forest, here's the bird species' that generally gives them instant ownership, if you tell them about the land, not just 'here's a piece of land, let's weed it'. So that's what Jim does. Gives you a history, gives you the passion behind Tathra Beach and why we should look after it.

On one of the working bees I joined, I had the pleasure of meeting Bruce, one of the founding members. Having run a nursery for many years, Bruce is exceptionally knowledgeable about local plants. Along the way he pointed out wattles that were endemic to the area and introduced me to various eucalypts. Without his knowledge, I wouldn't have appreciated the uniqueness of the plants we were caring for.

In the aftermath of the 2018 bushfires, which decimated parts of the town, there is often a conversation among members about how the bush is recovering and how long it is likely to take to recover further. There are also broader conversations about the role of a changing climate in causing more intense droughts that lead to bushfires. Unlike bushfire recovery agencies and organisations that are set up in the aftermath of the bushfires, the Tathra Land Care group is noted for its longer-term presence.

As I attend more working bees I can see a deeply embedded ethic of care for the land. The members draw satisfaction in being able to help, to make a difference, to do 'something that's worthwhile' and 'something good for the local environment'. Some members explained to me that even before moving to Tathra, they knew that they wanted to find ways to care for the environment.

> One of the things we said to each other when we chose Tathra to live, it was the spot that we'd chosen to settle in, and therefore felt a responsibility to give back in whatever way we could ... that's been fairly forefront in our minds for a long time.

Others explained that over time, they had come to appreciate the natural environment in Tathra and became particularly concerned with weeds after the Tathra bushfire. Pulling out bitou bush provides a 'feel-good factor':

> It gives me a little bit of power in a world where you haven't got much power ... I feel annoyed at that lack of leadership in terms of how we look after our beautiful resources, so this is a practical way of working locally.

For longer-term members, they keep coming along because working bees provide

> that satisfaction of pulling out a weed and knowing that doing so is making a difference. Finding them, and then pulling them out. And it always made me feel good to do, and it still does.

These days there is a real thrill in discovering a bitou bush plant. You may be happily walking and talking when all of a sudden, a member of the group excitedly shouts about having discovered a bitou bush. Depending on the size of the plant, others may be called over to help. A mature bush may be 2 m tall and 5 m wide, requiring multiple members to mattock their way to the heart of the trunk, and remove it by its roots. Plants are then hung nearby, roots off the ground, to stop them resprouting. There is a palpable sense of achievement among members that their dedication to the cause is still worthwhile after more than 30 years of effort. Another plant is added to the count of bitou bush removed this year.

Beyond getting to know and look after the environment, Tathra Land Care is a group in which friendships bloom and grow, where residents new to the area can find 'like-minded' people and where those affected by bushfires and other challenges can feel supported. The opportunity to chat to others while bushwalking with a purpose should not be overlooked. There is a sense of camaraderie that comes from crawling around the undergrowth and tramping through dense, steep country together; it creates a shared experience to draw on when casually meeting other members around town. Some relish being able to talk non-stop for three hours and know that there is a lot of emotional support from others in the group.

Stepping back in time

If you were to stroll along Tathra Beach today, you might wonder what all the fuss is about. The dunes are covered by a range of native plants: coastal beard heath (*Leucopogon parviflorus*), banksias and wattles. If you have an eye for weeds, you might spot some asparagus fern (*Asparagus virgatus*), bridal creeper (*Asparagus asparagoides*) or turkey rhubarb (*Rumex sagittatus*), but you would be hard pressed to find bitou bush.

Rewinding to the early 1990s, you couldn't go anywhere in Tathra without seeing this vibrant 1- to 2-m-tall green bushy plant with distinctive yellow flowers. A native plant of South Africa, it covered the dunes and the cliffs and was scattered in between the houses and vacant blocks. Local residents were concerned about the speed with which bitou bush was 'taking over the countryside' and 'destroying the character of Tathra'.[2] The local council agreed, describing bitou bush as 'the greatest pest plant in the Australian coastal environment, blocking access to beaches and destroying native bushland'.[3] And so the beach- and bush-loving Tathra community came together to care for this stretch of coast, nestled between Mimosa Rocks National Park to the north and Bournda National Park to the south.

But to really understand Tathra Land Care's epic eradication programme we need to go back in time to when bitou bush was first introduced to the area. In the 1960s, you would have found a very different dune system in Tathra. The beach was severely degraded and sand drifted freely across the road. The local community was concerned about the shifting sands and sought external advice about how the dunes could be restored and stabilised. The Tathra Beach Reserve Committee was advised by the NSW Soil Conservation Service that bitou bush should be planted. This unfortunately common recommendation at the time would come to threaten what residents and visitors hold dear about Tathra: its 'pristine' natural environment.

Planting bitou bush seed

Time and again throughout Australia's colonial history, we have sought to solve problems of our own making through the introduction of foreign plants and animals. In the world of weeds, one of the most famous cases involves prickly pear (*Opuntia* species) cactus. Prickly pear was widely planted by early settlers in the early to mid-1800s in NSW and Queensland as drought fodder. Within decades it had spread at an alarming rate, estimated at 400,000 ha per year, rendering much farmland unusable.[4] A 1930s video about 'the conquest of prickly pear' reports that 24 million ha had been smothered.[5]

The story of bitou bush is similar. During the early 1900s, coastal dunes along the NSW coast were being undermined through a combination of human activities: sand mining, use of vehicles for recreation and urban development.[6] As native vegetation was damaged or removed, beach and dune erosion became problems. Sand dunes shifted across roads and made access to public and private property difficult. A solution was needed and it came in the form of a plant from South Africa.

Bitou bush was inadvertently introduced to NSW in ships' ballast sometime in the late 1800s[7] to the early 1900s.[8,9] Observations about how easily it established in the 'inhospitable dune environment' around Newcastle, NSW, meant it was soon deemed a 'suitable species for sand drift control'.[10] From the early 1930s, bitou bush was taken from Newcastle and planted in coastal areas as a means of managing dune erosion.[11,12] In the 1950s, the NSW Soil Conservation Service wrote newspaper articles promoting the virtues of bitou bush as a suitable species for controlling sand drift because it 'grows rapidly in bare sand, developing into a considerable scrub' and is one of several 'outstanding' introduced plant species that are more successful at controlling sand drift than native plants.[13] This was what we might consider today to be the best available (colonial) science. So on the recommendation of a government department, in June 1964, a small area of bitou bush was fatefully planted near the Tathra Surf Club.

True to the description of bitou bush provided by the NSW Soil Conservation Service, the plant grew rapidly. It took only a few years for bitou bush to form a monoculture along much of Tathra Beach. By the early 1970s, it had spread to the headlands, drainage lines and hind dunes, crowding out native vegetation but doing little to stabilise the dunes.

While bitou bush was still two decades away from being declared a noxious weed in NSW, the Tathra community was concerned and began to take action. In 1973, the Bega Tathra Conservation Society began working bees with a view to eradicating bitou bush and revegetating with native plants.[14] Efforts were *ad hoc* and had no support from local authorities or the wider community because bitou bush was still largely valued for its mistaken potential to stabilise the dunes. So, despite the efforts of a few dedicated volunteers, bitou bush continued to flourish.

Concern about bitou bush spreads up the coast

During the 1980s, concern about the planting of bitou bush and its impact was growing, not only in Tathra but right up and down the length of the NSW coast.

Remarkably, a 24-person-strong bushwalking group spent 100 days in 1981 walking over 1,000 km from Tweed Heads in the far north of NSW to Sydney's Bondi Beach to survey the impact of coastal development and mining on 174 beaches. As part of this survey, the group documented that over 80 per cent of the beaches had 'significant infestations' of bitou bush. The group gathered over 18,000 signatures, which were presented to the then premier of NSW, Neville Wran, calling for an end to sand mining as well as funding for 'environmentally sound methods of eliminating bitou bush'.[15,16]

As concern built, councils and community organisations called for bitou bush to be declared a noxious weed in NSW, as had already happened in Victoria[17] and was poised to take place in Queensland.[18] The then NSW Department of Agriculture not only refused to declare bitou bush a noxious weed but it continued to advocate for its use as a dune stabiliser and was concerned that removal of the plant would lead to dune erosion. And so local communities began their own initiatives to eradicate bitou bush[19] and raise awareness.[20]

In 1984, a national conference was held in Port Macquarie, NSW, to 'improve exchange of information and to develop co-operative and supportive strategies against this species'.[21] This conference is seen as a turning point for bitou bush control nationally.

Incontrovertible evidence about the need to act in Tathra

At the same time as bitou bush was receiving growing attention across the state and nation, work continued in Tathra. The local council surveyed bitou bush in the area in 1983 and reported that it covered 200 ha.[22] In 1985, the Far South Coast Environment Group held a 'Bitou Bush Bash' which

> made the initial attack on the noxious plant to control its spread within the Tathra area. Volunteers worked for a week to clean up many areas of bitou bush plants, confining the infestation to a small area.[23]

Despite these sporadic efforts, it is estimated that by the early 1990s, bitou bush covered ~300 ha in and around Tathra. A hand-drawn map of Tathra records the extent of the bitou bush infestation. There were several 'heavy' infestations, as well as 'light and scattered infestations' throughout the township. A report from 1993 described the extent of bitou bush infestation at Tathra as taking up an area of '6–8 km^2 in and around Tathra Township.'[24]

By now the evidence was incontrovertible; planting bitou bush had been a big mistake. Any supposed benefits that bitou bush provided in terms of dune stabilisation were clearly dwarfed by the impacts it was having on native plants. The *Bega District News* reported that

> [a] creeping bush with lush, green foliage and brilliant yellow flowers has already established itself in many parts of Tathra.

> Thick infestations in the surf club corner, along the sand dunes and near the sewerage treatment works have overgrown and killed all native plant species.[25]

If left unmanaged, it was highly likely that bitou bush would continue to spread far and wide. A 1984 report on the distribution of bitou bush along the NSW coast[26] predicted that if bitou bush was left untreated in Tathra by 2010 it would spread as far north as Bermagui and as far south as Pambula, that is, ~30 km in each direction. Given that Tathra is surrounded by national parks, there were not only significant implications for the local community, but also for cultural values and ecosystems in the surrounding region.

The turning point in 1992

With such evidence at hand, it became easier to inspire the local community to act, to lobby governments to change the status of bitou bush and to successfully apply for funds for its control. And so multiple groups of concerned citizens began to coalesce around this issue in Tathra.

By the end of 1992, all the factors that would facilitate the establishment of Tathra Land Care (TLC) were in place. On 5 December 1992, Jim Kelly organised a 'bitou bash' at which 12 volunteers worked in pairs north along the beach from the caravan park to the first walkway. It took them three hours just to cover this 200 m stretch. Local residents Bill Caddey and Bruce Hamilton began to remove bitou bush from the Tathra Forest Reserve to the west of Tathra. At the same time, the Bega Environment Network applied for $150,000 JobSkills funding for a 'Bitou Bush Control Project', which was ultimately successful and would go on to employ up to 20 workers, 5 days a week for 6 months to systematically remove bitou bush from public land. The government agencies were also on board; Bega Valley Shire Council finally declared bitou bush a noxious plant in December 1992.

In March 1993, the three local community efforts coalesced to form a steering committee.[27] Together they distributed flyers with the heading 'News Flash: Bitou Bush takes over Tathra'. Their concern for the future of their local environment was clear, not only in the headline but in the blurb, as they explained that '[m]uch of the NSW coastline has already fallen victim to bitou bush, which now covers vast areas and is virtually uncontrollable'. But there was still hope.

> Fortunately, the bitou bush problem in Tathra (although threatening) is controllable and action is already being undertaken to reduce its spread. This is not an easy task and a *full community effort* is needed to strengthen the small band of dedicated locals who are trying to keep this noxious plant from ruining our fragile coastal environment.[28]

And so 'everyone who appreciates Tathra's environment and who doesn't want to see it ruined' was invited to attend a 'Bitou Bush Action Evening'. The last line of the invitation called for the community to '[a]ct now or we may regret our neglect'.

The call to action was effective. On 7 April 1993, 63 people attended the meeting at the Tathra Surf Club and heard about bitou bush, what was being done and how they could help with its control.[29] By some accounts, this meeting was when Tathra Land Care formally started. A few weeks later, on Saturday 24 April at 8 a.m., 18 residents, 'armed with machetes, tomahawks, bush saws and other implements of destruction' worked to remove bitou bush. Notes from the day record that bitou bush was 'easy to pull out in shallow rock and soil'[30] but there was the question of what to do with the piles of bitou bush removed and whether they should be burnt. There was also recognition of the need to cut bitou bush before it seeds in April/May, although bitou bush is known to flower intermittently year round.

Behind the scenes, the steering committee was working to build momentum, reaching out to other local community groups and developing a plan to achieve their aim. The steering committee contacted a wide range of organisations including the local council, Greening Australia, the Rotary Club, the football club, Tathra Beach Trust, Tathra Lions Club, Tathra Country Club and Tathra caravan parks. These groups were asked to help with the costs of buying herbicides, advertising group activities on community noticeboards, and for permission to do work on private property.

And so between April and May 1993, a formal structure and plan was coming together, and another public meeting was called for 23 June 1993 at the Tathra Surf Club. This would formally elect the executive of Tathra Land Care, allocate jobs and flesh out the plan. The membership fee was $5, and was still only $10 in 2022. The media release stated that this was the 'inaugural meeting of the 'Tathra residents against bitou bush Landcare Group"; it was originally intended that the acronym would be CRABB (Concerned Residents Against Bitou Bush), but ultimately Tathra Land Care, with the double acronym of Tender Loving Care, prevailed. In minutes Tathra Land Care (TLC) was 'voted in majority'. The group agreed that it wanted to be a 'non-financial' organisation because it didn't want to 'diffuse our energy by getting involved with money raising activities or financial accounting'. Officially, Tathra Land Care received its first 'start up Landcare grant, courtesy of the NSW Department of Conservation and Land Management' in August 1993.

Shortly after the second public meeting, the first of many Tathra Land Care newsletters was produced. In this first newsletter, the growing formalisation of the group is evident. It has a clear aim: 'The main purpose of TLC is to eradicate bitou bush from the Tathra Area'. There was a three-pronged approach to the control of bitou bush: 1) heavy infestations were allocated to council to spray sensitively; 2) moderate infestations would be addressed by the Bega Environment Network scheme as well as by monthly working bees from 8 a.m. to 11 a.m.; and 3) monitoring of small areas adopted by residents.

The importance of the relationship with the local council is evident in this strategy and has stood the test of time. The Senior Noxious Weeds Inspector attended the public meeting in April 1993 and supported the establishment of the group. He lent the council spray tank to the group on weekends to deal with heavy infestations. He also provided insights into council responsibilities for controlling weeds and made the most of changing legislation to support the group.

While Tathra Land Care was forming, legislation relating to weeds in NSW was changing and bitou bush was receiving increased government attention. The NSW *Noxious Weeds Act 1993* became law on 1 July 1993. The new Act sought to clearly identify which 'noxious' weeds require control, how they should be controlled, who has responsibility for controlling them and to provide a state-wide framework for their control. As part of this Act, four categories of weeds were specified depending on their level of spread and associated need for control. Bitou bush was declared a category 3 weed (W3) in the Bega Valley, meaning it was recognised as being well established.

In August 1993, local residents received a notice from the council that bitou bush was declared a W3 weed and that a maximum penalty of $4,000 was applicable to occupiers of land who do not 'implement reasonable and effective control, sufficient to at least prevent the spread and reduce its numbers and distribution'. Before sending this letter to residents, the council weeds inspector consulted with the Tathra Land Care steering committee. The final letter sent to residents in August 1993 echoed the words of the invitation to the public meeting. The council described bitou bush infestations in Tathra as 'although somewhat extensive are controllable with *all land occupiers working together.* Thus Council seeks your support in controlling this weed on land you occupy.'[31]

Beyond working with the local council to promote their campaign, the group was highly active in engaging local media. In 1993 alone, seven articles about bitou bush and group activities were published in the *Bega District News*. The group also sent an article to the *Goulburn Post* and made calls to *ABC Radio* to broadcast community service announcements about upcoming working bees. This media campaign was supported by the Bega Environment Network, who distributed their own brochure about bitou bush to the local community titled 'Help Save Tathra from the Bitou Bush Threat'.

In August 1993, just as Tathra residents were ramping up their efforts, the then NSW Department of Agriculture produced the inaugural *Bitou Bush and Boneseed Bulletin*. From this we see that while Tathra Land Care was holding its first steering committee meeting in April, a second National Workshop on bitou bush was being held in Port Macquarie bringing together scientists and government officers. Notes about the national workshop reveal how pivotal communities were to the monitoring and control of bitou bush along the NSW coast.[32] Of particular note is a letter (page 5) from Dr Mark Lonsdale, CSIRO Division of Entomology, addressed to 'all community volunteers involved in bitou bush'. He noted that

> the public are strongly committed to doing something about bitou bush. We all owe a debt to the various community volunteer groups that have carried out such fine work to reclaim dunes from bitou.

Throughout the rest of 1993 monthly working bees continued and there were many spraying activities in between to support these efforts. A promotional video produced by the group shows footage of the steep cliff between the wharf and the beach where Tathra Land Care had been spraying. For this difficult work, the group recognises the loan of the council spray rig. The video also shows other areas that had been treated by the Bega Environment Network JobSkills programme.

Achievements were celebrated at the end-of-year meeting. Perhaps prematurely, it was declared in a Bega Environment Network newsletter that '1993 has broken the back of bitou in Tathra'.[33] The same enthusiasm is evident in the promotional video, where one member stated that 'we've probably eliminated from the Tathra area around half the bitou bush' and predicted that 'within the next 5, 6 maybe years, we will have a bitou bush free community and it will be a much better place for us to live'.

The belief that eradication of bitou bush, like other weeds, could be achieved over a relatively short period of time was persistent. Even in 1996, a report produced by the Local Landcare coordinator expresses confidence that '[g]iven the cooperation of private land owners, all mature plants in Tathra should be prevented from setting seed by the winter of 1997'.[34] With the benefit of hindsight, we know that while the concentrated efforts to control bitou bush in Tathra in 1993 were important in breaking the back of the infestation, a further 30 years of consistent and systematic effort would be required to eradicate the plant almost entirely.

Keys to success

It is no mean feat to sustain a voluntary group for over 30 years, especially when it involves weeding, something many of us would rather not do in our own gardens, let alone on public land.

The group has benefited from consistent and passionate leadership for the length of its duration. It has received financial support as needed and has developed a culture where there is a role for everyone and everyone is valued for the time that they can contribute.

There are two men who shared leadership of the group for almost 30 years: Jim and Dave. Between the two of them they would phone the full membership list every month to see who would be attending, providing insights into who was unwell and who was away, and would actively recruit new members. According to those present today, it is almost impossible to move to Tathra without being invited to a working bee. Members speak of how they were motivated by the dedication of Dave and Jim and other group members; it made them want to 'do my bit too'.

The commitment of the group not to 'depend on government grants' and not 'dilute our efforts by fundraising' hides the help that they have received along the way and the work that members must have been doing in the background to keep the group going. Aside from the initial grant from the NSW Department of Conservation and Land Management in 1993, the group received small amounts ($1,800 to $6,000) of funding over time that was used for equipment and herbicides.[35]

The relaxed nature of the group, where a conscious effort has been made to keep administration to a minimum and where each member is valued for any time they are able to contribute has been key; some people would join for a meeting or two, others would stay the course for a decade, and there are others who have been members for over 30 years.

Over time, the leadership team have continued to motivate members by frequently reflecting on what they've achieved. In the 2001 video, you can hear Jim's voice behind the camera asking each volunteer about how many bitou they'd found and whether they believed they were making a difference. One volunteer responded that 'I think if we hadn't started this would be wall-to-wall bitou by now and as it is we've got to hunt for it now'. The quarterly newsletters reinforce the success, with reports on ever-diminishing numbers of bitou bush found at the previous working bees.

The presence of longer-term members also contributes to the collective memory and success story. Their eyes light up when talking about 'how Tathra has been transformed through the efforts of volunteers' and how 'behind the wharf, that iconic structure there, the backdrop was bitou bush. It's not now. It's got a nice hue to it with the native bush.' They recall that when they first started in the 1990s, 'when we started tallying them up [bitou bushes pulled out] at the end of a working bee and it wasn't even a hundred, there were thousands ... it wasn't uncommon to have two hundred each'. And now, there are working bees when only one or two are found. Newer members can see that 'there's a real sense of pride in what they have done, because they have changed the landscape, they have re-established the native options in the coastal areas and the natural environment'.

The sense of achievement and progress has helped sustain the group. Not only is it recognised within the group but also by those in the broader community, government and non-government organisations alike. Other residents have described it as a 'very successful project' because they haven't seen any 'bitou bush for the last few years'. Another described the concerted effort as 'pretty extraordinary'.

The council is equally enthusiastic describing it as 'one of the best ones ever ... it's an amazing outcome for something that could have been astronomically ... expensive. It would have stretched up and down our coast.' Similarly, the Far South Coast Landcare Association recognises the value of the group:

> [W]hat a massive difference they've made to this area ... If they hadn't started, kicked off 20 years ago, I don't think we could get on top of bitou now.

Beyond recognition from local organisations, the group has won several regional awards. Nationally, Tathra Land Care is one of only two case studies of successful community control included in the national *Bitou Bush Management Manual*.[36]

Beyond working bees

There is widespread agreement that Tathra Land Care working bees have been instrumental in reducing the extent of bitou through their ability to deliver on the long-term, repetitive task of scouring the landscape for bitou. And yet if we only describe the group's working bees, then we miss the other activities that they have undertaken over the years, as well as the complementary projects and other people and organisations who have supported the group's efforts and have been key to keeping the momentum going.

While the working bees are key, the group has engaged in a range of activities to control bitou bush. In the early years the group had four objectives that would help them to achieve their aim of eradicating bitou bush. These included:

> 1) Destroying all mature, flowering plants so that the production and dispersal of seed would be stopped.
> 2) Locating and mapping bitou bush infestations in order to systematically use our resources to control bitou bush
> 3) Making a commitment that to destroy re-generating bitou bush would require work over a 7 to 8 year period.
> 4) Eventually rehabilitating areas of heavy bitou bush infestation by planting natives.

The first two objectives were deemed completed by the end of 1995 and 1993, respectively. The creation of a map of infestations in 1993 was particularly significant. This map formed the basis for future working bees, where and when work would be undertaken. The fourth objective was occasionally undertaken through planting days. For example, tree planting first appears on one of the work maps as an activity in 2010 and then again in 2011.

Most of the effort in the intervening period has been on destroying regenerating bitou bush. There were three ways the group achieved this:

> 1) Working bees – where plants are pulled out by hand, dug out with a mattock, cut with secateurs and then herbicide applied to the stump of larger plants.

> 2) Spraying programme – use of the local council spray tank and herbicide that had a pump and 200-m-long hose. This was used for larger infestations and those in difficult to reach areas.
> 3) Small area contract scheme – where volunteers take responsibility for a small area of Tathra to keep an eye out for bitou and record work that needed to be done.

The spray programme was mostly undertaken by volunteers in the early days, with group members operating the council spray unit for 3 to 4 hours on the remaining three Saturdays a month. Since then, council and contractors have done the spraying work. For the first decade, such spraying was a regular feature of the programme before becoming more sporadic. 2007 is the last year that spraying appears on the work maps.

The small area contract scheme was short lived. In this scheme, Tathra was divided into at least 25 areas, which residents were encouraged to adopt and take responsibility for monitoring for bitou bush. The idea was that it would enable people to contribute who may not be able to attend working bees or who prefer not to be involved in the spraying of glyphosate.[37] Only one map from 1993 reveals who had signed up to each area. The scheme was mentioned in a 1996 article about the group's activities; it was estimated that the small area scheme contributed 100 hours of volunteer work per year.

Instead of the small area scheme, what evolved was that some people volunteered to help survey areas for bitou bush that were not reached during working bees. These efforts were meaningful for local members, who appreciated that work was done in between working bees: 'Jim and Dave often tell us that during the intermediate period during the working bees, so-and-so went and pulled out hundreds on one of the sand islands of the river.' These individual and small group efforts have been a feature for much of the history of the group but receive little mention in group documentation.

While not mentioned in the list of objectives or activities, a document from 1996 records the group's diverse public awareness programme. It has included a photographic display of bitou bush exhibited for a month at a time in the public foyer of the council chambers, the Bega Environment Network Centre, a display at Candelo markets, ~35 km to the west of Tathra, at the local primary and high schools, the Tathra Post Office and the NSW Department of Conservation and Land Management stall at the Bega Show. They still produce a quarterly Tathra Land Care newsletter, submit articles to the local newspaper and do interviews on local ABC radio. There are still two prominent Tathra Land Care signs as you enter Tathra from the west and north that inform the community and visitors about bitou bush control work being done.

Not mentioned anywhere but evident through the group documentation is the importance of keeping records. The group has copies of the newspaper articles published over the years, and a map for each year, documenting where activities were taking place. There are two videos of group members at work, as well as copies of the various resources that the group has produced over time. Without such documentation it would be hard to understand the multiple endeavours that it takes to run a long-term weed eradication campaign.

Broader community support and efforts

Several other community organisations made important contributions to dealing with bitou bush in Tathra. The Bega Environment Network's JobSkills programme played a key role in 1993, when Tathra Land Care was finding its feet. During the early 1990s, The Bega Valley Tree Planters (previously named the Bega Valley Men of the Trees) responded to calls from Tathra Land Care to conservation groups to come and help with their working bees. In 2002, Green Corp helped to cover much of the headland and some of the area near the sewerage works, and in 2004, Australian Volunteers Abroad helped with a similar area.

Such initiatives were aided by the formal declaration of bitou bush as a Weed of National Significance in 1999. As part of this declaration, a National Plan for bitou bush was developed. The initial plan identified a northern containment line at the border with Queensland and southern containment line at Tuross Head, almost 100 km north of Tathra. The lines were drawn on the expectation that any infestations of bitou bush beyond the national containment zone were eradicable. Thus, less than 10 years from the establishment of Tathra Land Care, their future success was already anticipated.

A *NSW Bitou Bush Strategy*[38] and a *South Coast Regional Bitou Bush Strategy*[39] published in 2001 and 2002, respectively, reinforced the important role of community groups in controlling bitou bush. Although no specific groups were mentioned, a newspaper article kept by the group records that they were recognised by the *South Coast Bitou Regional Bush Strategy* as an advanced demonstration site for bitou bush control.

These supporting activities show that Tathra Land Care not only directly achieved bitou bush control but was also an effective community organisation that harnessed the support of numerous government and non-government initiatives.

What now?

As the group reflects on what it has achieved over 30 years, the swathes of bitou bush it has removed, the knowledge it has shared, the challenges it has overcome, and the friendships made, a new question is beginning to emerge … what now? How vigilant does the group need to be to ensure the few remaining bitou plants do not become re-established? How can the younger generations be encouraged to help look after Tathra?

Some within and beyond the group wonder whether the remit of the group should be expanded to include other weeds. In the final report for the JobSkills project in 1993 it was noted that '[w]hile Bitou was our target species, there are many other species of potentially invasive weeds in Tathra bushland that are known to progressively degrade native ecosystems, increase erosion and interfere with drainage'. The report provided a list of 14 other weeds, many of which are the focus of Landcare groups across NSW and Australia, such as asparagus fern, lantana (*Lantana camara*) and privet.

Members recognise that if they were to control another weed, it would need to be something achievable. Their own experiences have shown how a local eradication campaign needs more than enthusiastic members. The help of multiple organisations and the commitment of local council is key.

While having a largely singular focus has undoubtedly helped the group to achieve its ultimate aim of eradicating bitou bush from Tathra, members know that for some, 'just looking for bitou bush all the time can get a bit boring', and that it's probably a reason why 'people who are new probably don't keep coming back'.

Instead of starting to control other weeds to overcome the monotony of searching for bitou bush, some members would like to see more diverse tasks included in the schedule of activities, such as planting trees and other native vegetation. While the group has had the occasional planting day, it is far from a regular occurrence. Suggestions include reducing the frequency of the bitou bush working days to once every two months and then including more diverse activities the other months. This may include meetings with a more educational focus, where local experts are brought in to talk about bushfire recovery and the local plants, which is particularly appealing to those who are newer to the region and keen to learn more about where they live.

8

Coming together to clean up Candelo Creek

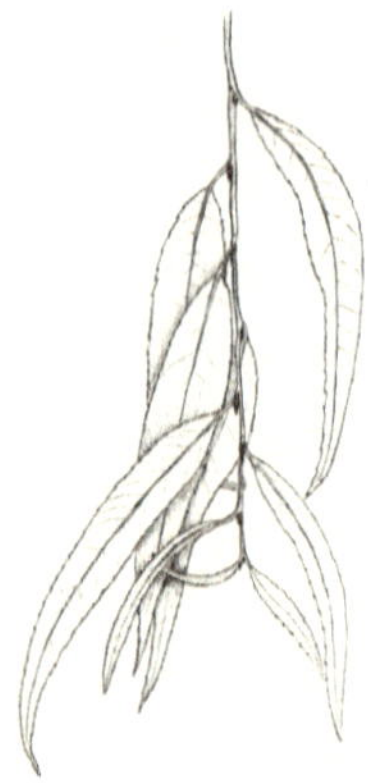

Candelo Creek is at the heart of the town – its geography and history. On one side of the creek is the General Store, cafe, post office, park, tennis courts and primary school; on the other the Candelo Hotel, pub, bowling green and swimming pool. The valley sides rise steeply around the township, enclosing the creek. This means that no matter where you are in Candelo, whether you're buying your groceries, having a coffee with friends or a drink at the pub, you can hear the creek running, the frogs croaking and the birds coming to feed and chat in the trees that line the riverbed.

The need to cross the sandy river is part of the foundation story of the town. Some say the name came from wagon drivers needing to call for a candle to light the way across the river.[1] A fictional book by Georgia Blain,[2] named *Candelo*, describes it thus:

> *They were on horseback*, she said. *There were no streetlights. They had no torches.* And as they clattered over the bridge, they called out for light. Candel-o, their voices ringing out in the darkness.[3]

Back then, in the late 1800s, the creek looked nothing like it does today. The land had been cleared and only bare soil remained. Photos from the time reveal a sandy and silty watercourse, with barely a tree in sight.

Not long after the town was established, the new residents began to plant trees. The ones planted closest to the waterline were willows. Experience from England showed that these trees were good to plant in wet places and could also be used as a source of fodder for cattle. The willows thrived along Candelo Creek. Residents and travellers alike commented on the willows, the shade they provided for social events, and the English feel they gave to the countryside.

Fast forward to today, and there are no longer any willows in the stretch of Candelo Creek that runs through the centre of town. They were removed by the Landcare group in the early 2000s in recognition of the detrimental effects they were having on the ecological integrity of the river system.

So what happened in the intervening years? How did willows go from being prized to problematic? How has the community responded to the changed landscape that has come with removal of willows?

The answers to these questions require an understanding of how willows have been viewed in Australia since colonisation and how the proliferation of willows in Candelo and the Bega Valley more broadly set off alarm bells about their impact.

Willows in Australia

The introduction of willows to Australia can be traced to the early 1800s, when Governor King, the third governor of NSW, declared that willows would be among the trees to be generally planted.[4]

This positive view of willows dominated for most of the next 200 years. People were drawn to their aesthetic and hardiness. In the late 1800s, government botanist Baron Ferdinand von Mueller noted the many virtues of willows, including their resilience to bushfires.[5] In the early 1900s, as the Director of the Royal Botanic Gardens Melbourne, Victoria, he valued willows for preventing evaporation from rivers.[6] Into the 1950s, Australian gardening books advocated for willows because of their generous shade and ability to prevent soil erosion.[7]

Beyond their values as trees, white willows (*Salix alba* var. *caerulea*) were also valued for their wood. Many sports articles, historically and currently, refer to willows in describing cricketers. Because cricket bats are made from willows, references are regularly made to 'willow wielders' to describe those batters with impressive skills and run scores. While most cricket bats are imported from England, a plantation of willows in Daylesford, Victoria, was used to make cricket bats from the early 1900s to the 1950s.[8] While some bats are still made there today, the early to mid-1900s was clearly the heyday of willows in Australia.

John Dwyer[9] situates this passion for willows in Australia within a global context. He points to the prevalence of willows in Claude Monet's paintings as well as in Shakespeare's texts, among other significant cultural references.

Yet despite the colonial passion for and cultural significance of willows, environmentalists in Australia have become increasingly concerned about willows over the last four decades.

Initially, it was believed that willows introduced to Australia were not able to reproduce and so their location would be determined only by where they were intentionally planted. This assumption was shown to be incorrect in the 1990s when willows began to hybridise.[10]

The way in which willow numbers multiplied in the Bega Valley was of particular concern on the national stage, drawing attention of the *Sydney Morning Herald*:

> War has been declared on the willow as its exploding numbers continue to damage NSW waterways. The State Government has been forced to act and has asked its Noxious Weeds Advisory Committee to investigate whether some species should be classed as noxious and removed ... One of the worst affected is the Bega River system where, over the past 150 years, the number of willows has grown from 2000 to more than 2 million ... In Bega, the [NSW] Department of Land and Water Conservation said infestations of willows had eroded the riverbanks and caused flooding in some places.[11]

From this time onwards, public concern about the impacts of willows grew and with it came investment in scientific studies. Research was undertaken into the impact of willows on streamflow, erosion and habitat.[12] As the body of research grew, it became increasingly evident that willows degraded riverbanks, reduced the quality and flow of water, limited habitat for fish, birds and insects, and obstructed recreational access to rivers.[13] Like bitou bush (Chapter 7), willows had once been heralded as a species that should be planted to prevent erosion but were later found to exacerbate soil loss and impact stream hydrology.[14]

What is perhaps surprising is just how quickly willows fell out of favour. After almost two centuries of being a prized plant, their downfall, at least among those in environmental circles, was swift.

In 1999, willows were listed as a Weed of National Significance in the inaugural list of 20 weed species. Since then, much research has been invested into understanding the best ways to prevent the spread of willows and much funding has been allocated to willow removal.[15]

Yet there are still some who speak in favour of willows. Notably, Peter Andrews OAM, who developed the principles of Natural Sequence Farming, saw willows and other weeds as useful in assisting with land rehabilitation.[16,17] Retired Queen's Council and former chairman of the Australian Garden History Society, John Dwyer,[18] is also in favour of willows. He argues that willows have helped to prevent stream-bank erosion and that they are still valued for their aesthetic in the landscape. He argues that '[w]illows have a profound cultural resonance ... [and] there is still a place for willows in our gardens and cultural landscapes'. Indeed, a novel titled *Willowman*, published in 2022,[19] can be considered an ode to willows, the people who work with them to make cricket bats, and the cricketers who wield them. This continued passion for willows means that the debate about willows on the national stage is yet to be settled.

Caring for Candelo Creek

The case of willows in Candelo provides a concrete example of how our relationship with willows has changed over 200 years and the difficulties that Landcare groups face in navigating changing public opinions.

Dozens of newspaper articles from the late 1800s and early 1900s refer to willows in Candelo, the nearby Kameruka estate, and slightly further afield in Bega, Cobargo and Bombala. The way willows are described in these articles indicates the main value of willows was for their aesthetic and as a social setting for the local community.

> Kameruka House is built on a commanding position, overlooking many miles of country. It is situated at the junction of the Candelo Creek with the Tantawangalo River. Bordering these streams, for some distance approaching the junction, are a row of stately willow trees, perhaps the finest plantation of willows in the colony. They half conceal the river, over which they spread their streaming foliage.[20]

Local newspaper articles from the early to mid-1900s describe how diverse social activities – Sunday school picnics, post-golf afternoon teas and stockmen's lunches – were held in the shade of the willows. These articles reveal that willows were far from a nondescript green backdrop to social activities; willows facilitated these events and were appreciated for doing so.

It is unclear when willows were first planted in Candelo. While references to the shade of willows indicate that mature trees already existed in the late 1800s,[21] newspaper articles only reported on willow planting events from 1907 onwards: the 'local Progress Association has been instrumental in having willow trees planted on the creek side, from the showground beyond the traffic bridge'.[22] Similar tree planting events were reported in 1909 and 1912 as a means to 'beautify the town'.[23]

Later, willows were planted for use as stock fodder during dry conditions and as a form of erosion control to minimise the effects of floods. Articles around the time of the 1918–1920 drought commented on the usefulness of willows as fodder: 'The highly nutritive properties of the common willow have become fully realised.'[24] After the drought broke, articles continued to highlight the benefits of willows for stock and advocated their planting.

> Now is the time to plant willow trees. During last summer and autumn the foliage of these trees provided invaluable fodder in this district. Willows will grow anywhere along the banks and

> gullies about Cobargo. The roots prevent erosion of the banks, and the trees throw a fine shade. Every farm should have willows along its water courses.[25]

The valuing of willows as stock feed and shade marked a change in the quantity of willows being planted. Where previously willows were planted in small numbers, now hundreds of them were being planted along water courses.[26] Indeed, one local councillor planted 900 willows in nearby Bemboka to feed his herd of 100 cows.[27]

After the drought, not only did newspaper articles promote the agricultural virtues of willows but it became a government requirement that lessees plant willows.

> The lesee [sic] shall within one year from the date of notification in the Government Gazette of the grant of the lease, place a row of willow trees along the frontage to Candelo Creek so as to be effective against erosion by flood.[28]

This requirement points to another value that was being attributed to willows; they were seen to be beneficial in holding banks together and reducing the impacts of floods. Newspaper articles about Bega, Cobargo and Dignams Creek mentioned willows being planted for erosion control. Closer to Candelo, willows were planted along the road to prevent erosion.[29]

At the same time as articles were being published about the need to plant willows, others expressed grief and anger at damage to and loss of willows in Candelo. Dating back at least as far as 1916, residents voiced concerns about willows being vandalised.

> Complaints are being made about destruction of the willows growing along Candelo river in the town area ... It is apparently the work of mischievous boys. Citizens with an eye to the beauty of the place went to a great deal of trouble in planting these trees, and it is a shame they should be destroyed in this way ... if the culprit is caught citizens who take pride in these beautiful trees will certainly take steps to make an example of him.[30]

A similar article appeared a year later, again with the threat that if the 'brats' were caught then 'there will be something doing'.[31] Aside from wayward youth, the other theory about who was cutting trees were stockmen, who were believed to be cutting trees 'for the purpose of feeding stock'.[32]

This concern for willows continued into the 1930s, when newspaper articles warned 'against the unauthorised lopping of the willows growing in Candelo Creek. These beautiful trees ... should be rigidly protected.'[33]

Beyond willow vandalism, the loss of willows is mentioned in descriptions of the damage caused by the river flooding. For example, in late February 1919, a flood that washed away Candelo's footbridge was noted for its impacts on willows: 'Most of the beautiful willows on the river toppled over and were carried away.'

This context is important for understanding why willows were once valued in Candelo, the emotional attachment residents had to willows and the landscape aesthetic they provided, and the significance of willows being deemed a weed in this place.

In all this time, there were only two people in the district who reportedly questioned the virtues of willows. In 1932, two councillors voted against a motion that would allow 600 willow trees to be planted in the Bega River.[34] They were concerned that the willows would affect the flow of the river, lead to erosion and affect adjoining lands. Ultimately, approval was granted for the planting to proceed, based on the votes of the three remaining councillors.[35] The two voices were no match for the tide of appreciation for willows. It would take another 50 years before the tide would turn.

A 1989 heritage study of Candelo perhaps marks the beginning of a broader shift in the way willows were perceived in the area. Generally, the study saw willows as unproblematic. Willows were noted as providing 'an important gateway entry to the township' that 'provides integrity and definition to both creek/township and pastoral landscape types'.[36] However, mention of the need to remove young willows and the value of cattle and horse grazing in preventing the spread of willows, indicates emerging concern about the impacts of willows on the 'recreational visual resource of the creek'.

Only four years later, a mass seeding event of willows in the district would turn this sense of unease into alarm.

> A willows expert and honorary research fellow at the CSIRO, Mr Kurt Cremer, said more than a million seedlings became established in the Bega River in 1993, growing as high as three metres in as many years. 'We still have a window of opportunity to control the trees, but we must act now,' Mr Cremer said. 'We have only recently become aware of the problem so it is only now that we can do something about it.' The [NSW] Noxious Weeds Advisory Committee will consider placing restrictions on the propagation and sale of willow trees and the removal of species in areas where they are an environmental threat.[37]

Weeding willows in Candelo

When Landcare started in Candelo is a little vague. Documentary records indicate Candelo and District Landcare received almost $17,000 in funding from the National Heritage Trust in 1997–1998 for a project on grassland decline. Yet the first mention of the group working on weeds in the creek dates to 2003, with just over $17,000 in funding received from the Commonwealth government's Envirofund for 'Rehabilitation and Enhancement of Candelo Creek':

> The project will undertake the strategic removal of willows, blackberry and African lovegrass from the streambed and banks. Small pots of native vegetation will be planted in place of removed weeds when conditions suit. Indigenous floristic restoration is the first step in the riparian recovery of this highly degraded section of Candelo Creek. Aquatic habitat restoration through pool reforming and stabilisation is the next step. Project outcomes will include bank and bed stabilisation, reduction of seed source for willows and other weeds,

> stabilisation of sediment, and demonstration of the landcare ethic.[38]

However, this project was not the first to tackle willows and attempt to restore Candelo Creek through tree planting. A clipping of a local newspaper article kept by the group makes mention of Candelo Rivercare tackling weeds and planting trees in 1995, to achieve an overall aim of restoring the creek to pre-clearing conditions. Landcare and Environmental Action Program (LEAP) Trainees reportedly helped the group pull out young willows growing in the creek bed. Bega Valley Tree Planters (previously known as the Bega Valley Men of the Trees) also provided volunteers to help at a tree planting event.

In interviews, long-term residents explained that initially Candelo Rivercare and Candelo and District Landcare were established separately because 'primary producers didn't want to be pushed around by those greenies'. There was clearly conflict over how the creek should be managed.

In 2002, a public meeting was held to discuss the problem of weeds in the creek. David Barton was voted chair of the Landcare group that would focus on eradicating weeds and planting native riparian species. Crack willow (*Salix fragilis* var. *fragilis*) was singled out as being one of the worst weeds the group needed to address. As one member explained:

> I think David Barton, he and Cliff Massey, I think, organised a community meeting on how to deal with the whole river system, Candelo Creek system. The farmers all got upset ... took off and left.

Those who chose to be members of Candelo and District Landcare decided to seek funding to control weeds on public land, particularly within the riparian zone, and volunteers were enlisted to 'help chainsaw, chop and set fire to the piles of willow'.[39] As one member explained:

> We had two aims initially, which I think we're still working on, which was to clear out the willow, but now that that's been done, we're just still keeping the willow out as much as possible.

> So, that was our main weed. And to improve access for the community to water.

Working bees in the early days were held monthly on the first Saturday of the month from 9 a.m. to noon, followed by lunch.[40] Later they moved to 10 a.m. each Friday and now are held every Thursday morning.

To support their work, the group has received funding from various sources, including the Envirofund grant mentioned above and the NSW Department of Infrastructure, Planning and Natural Resources, which they used to pay Koori[41] Revegetation Contractors to help remove willows.[42,43] A further $42,822 was received from Envirofund in 2006 to continue these activities.[44] Overall, the group has received around $200,000 over the 20+ years of the group's operation.[45]

Reports on how the funds were used indicate that Candelo and District Landcare, like Upper Kangaroo Valley Landcare (Chapter 4) and Deua Rivercare (Chapter 6), took a staged approach. They started first with the area downstream of the traffic bridge, then slowly made their way upstream. The group was so effective in removing willows that by 2009 there were none left in the stretch of Candelo Creek within the township.

> We started the willow control in 2003, and it's over a kilometre stretch of creek, and it was all done by 2009. We just gradually progressed up, killing the willows, chopping them down, and burning them in piles. Replanting, with a series of pretty hefty grants. People could see the work.

But anyone frustrated with weeds will tell you, it's not a job you can ever really finish, no matter how much time or money you have at your disposal. Besides keeping new willow seedlings in check, there were plenty of other weeds on which the group continued to focus. These weeds continued to be tackled by regular working bees, comprising local Landcare members with the occasional support of supervised volunteers from Centrelink, the Green Army[46] and GreenCorp.[47]

Weedy working bees

Every Thursday morning from March to November, members of Candelo and District Landcare congregate to tackle weeds somewhere along the 1 km stretch of creek that runs through the length of town. There is no shortage of weeds to choose from: stinking Roger (*Tagetes minuta*), blackberry (*Rubus fruticosus*), fireweed (*Senecio madagascariensis*) and hemlock (*Conium maculatum*), just to name a few.

From 10 a.m., the group slowly and methodically deals with the weeds along a patch of the creek.

One time I joined the group, the plan for the working bee was to burn off weeds that they had cut and piled up previously.

Eight of us met on the creek bank between the pub and the old bowls club. Philippa was the first. As the others arrived, she pulled out a 3D model of the creek that she had made from old tree guards. She made it to show the local school kids how the sand builds up along the creek, but it was also fascinating for the Landcare members. It sparked a conversation about the creek and how much members value it. They love swimming in it, taking their dogs for a walk along the creek, watching kids playing in the water and playing on the banks. They appreciate it just being there: 'What makes me feel good is when I find kids down in the creek, just building cubby houses, getting creative, writing poetry down there!'

Nearby, there were three piles of dry weeds, each up to a metre tall. I was excited to be doing something other than weeding or planting trees. This time the tools we needed weren't just our gloves and herbicide, it was our buckets and water packs. A retired member of the Rural Fire Service was there to help and make sure the fires stayed under control.

We started by clearing nearby vegetation so that it wouldn't catch alight. We filled the buckets and packs with water from the creek before setting fire to the piles. Some were tasked with raking and tending to the fires, one watered the plants from the latest tree planting, while the rest of us tackled the weeds, adding them to the piles as they burned. The crackle of the fires and the croaking of the frogs provided our soundtrack.

The blackberry made itself felt, its thorns piercing my gloves. And the stinking Roger seemed curiously named, it gave off a rather pleasant smell when burned. There was the occasional fireweed, which is considered one of the worst weeds in the district, mentioned by all interviewees as the key weed of concern to them on their properties.

No matter where I looked, there were more weeds to be pulled, cut and burned. But I didn't see a single willow, their absence testament to the group's success.

Beyond the satisfaction of weeding and seeing a willow-free creek, what was fun about the working bee was being in the riparian zone. Making our way through the sandy creek bed, noticing the plants – their location and which ones grew together, feeling the crunchy ground moving underfoot, and finding spots deep enough to fill the buckets. As one member explained: 'I just like being out in it and pulling up weeds. Or walking along the creek and taking a photo and posting it!'

For those who come regularly, this interaction with the creek is part of the thrill.

> I really enjoy actually seeing just how the creek's changing. It's really exciting to think that the willows, once the willows were gone, then this historic load of sediment is clearing out, and it's just a really fascinating process to see it.

We continued working for three hours. Slowly, we stopped adding to the piles, leaving the remaining weeds for another day. The sound of sizzling embers then gushing water dousing the flames signified the end of the event. There was no formal socialising afterwards, but some of us did make our way to the Candelo Café for some much-deserved coffee and cake.

Rehabilitation and education

Candelo and District Landcare is about more than weed removal. Their projects have involved rehabilitation and education as well.

As one member explained, the group is about

> [t]he successful restoration of a 2-kilometre stretch of Candelo Creek, where it runs through town. I don't want to use the word 'restoration'. I think 'rehabilitation' is better. When you're rehabilitated, you don't go back to what you were before. The creek can't go back to what it was before. The morphological circumstances are now different. You can't turn back time and make it better, but what you can do is, you can control the banks with vegetation, you can control the sand upstream and stop erosion happening upstream.

To achieve such rehabilitation, the group has held many tree planting events. In the first two-year phase 500 trees were planted, with a further 500 planted in the second phase. In the 20 years since then, there have been semi-regular tree planting events.

National environmental days, such as Clean Up Australia Day,[48] National Tree Day[49] and World Environment Day,[50] have provided opportunities to engage the broader community in tree planting and rehabilitation events. Much of the funding that they have received has been for this purpose.

As with weeds, planting trees is only successful if it is part of a sustained effort. Research on tree planting success shows that pre- and post-planting weed control is key.[51] If all you want to do is turn up, dig a hole, plant a tree and walk away, you might as well not bother.

So here is where the weekly working bees are important. They provide a mechanism through which seedlings are looked after and those that die can be replanted.

> Once you have a project, you've got to maintain it! And everyone's excited when it's happening, but the 20-year maintenance program is boring! ... You can do stuff at the start, but when you're working with the natural environment you've got to come back every season.

The success of tree planting for those in the group is about more than the number of people who turn up to plant trees, or the number of trees that are planted. It is about seeing the trees mature and begin to reproduce on their own.

> And the other thing that's happening now, here, with this particular project, is that ... succession is starting to happen. With a 20-year project, we're starting to get some succession.

The first time I joined Candelo and District Landcare for a working bee, the creek bank was awash with activity. There was a contractor explaining the best way to plant trees to a gaggle of school-aged children and their parents. The holes had already been dug and the trees and shrubs intended for them waiting alongside. All that was required was for people to pick a hole, place the seedling in the hole and cover the roots with soil, then place a tree guard gently around it. Together, 400 lomandras, native shrubs and trees were planted on the eastern bank, just downstream of the bridge.

But this event was about more than tree planting. As with many of the group's activities, learning from the past about the creek and the surrounds was front and centre.

The event had started with a yarn by a Yuin elder, who shared insights into the connection that Yuin people have with this landscape.[52] Across the road, indoors, women were learning how to weave from another Yuin elder. Such activities provided a more inclusive space, where those who were not able to join the tree planting were able to be actively involved in the day.

A Landcare tent was set up on the grassy area near the playground. It housed an information board that explained the importance of the creek and what the Landcare group had done to support it for more than two decades. Photos on the pinboard showed similar tree planting events held over time. They also showed what the creek looked like in the early 2000s as a point of comparison to today.

The way this day was organised reflected one of the group's aims: to educate the community about the value of the river and the need to look after it.

Setbacks

A surprising theme during the interviews about Candelo and District Landcare was the way in which their efforts were inadvertently undermined by staff working for, or contracted by, the local council.

Overall, members were enthusiastic about the council and the support that they had received over the years.

> We've recently got quite good support from council. It's really key to have those other people. You can't just be a group of volunteers on your own, not unless you are a much bigger organisation than we are.

Beyond being supportive of the group's activities, the council has provided funding to the Landcare group over much of its 20-something years of operation. This included providing grants, such as $5,840 from the Community Grants programme in 2014[53] and $3,500 for the Queen's Jubilee planting in 2022,[54] as well as funding contractors to support the weed control activities of the group.

Yet on three separate occasions the council has inadvertently undone work done by volunteers.

The first incident described by a member pre-dates the Landcare group. It involved an occasion where the council ripped up newly planted trees in their efforts to align the creek bank.

> One of the last things that the Bega Valley Tree Planters did was a tree planting on the Candelo Creek ... a year or so after, the council sent a contractor down with a backhoe or something, and they were rearranging the bank, and they obliterated most of the tree planting, which the council did provide money to replant.

Sometime later, when trees the Landcare group had planted as seedlings had matured, some work undertaken by council cut down the trees.

> we had an active grant going when the Council engineers, the road traffic engineers ... said, 'the sight lines as you come off

> the bridge are bad.' So, the council traffic engineers sent in a crew, and they just chopped down a whole lot of eucalypts. So, it kind of makes you feel like, we were here. Why didn't somebody come and talk to us?

Most recently, a third tree planting was undermined by council.

> A few years back, probably three years ago, one of their plantings right in town was demolished by the council town team. That sort of stuff just guts a group like that. That just go, 'wow, how can we feel we've got this really good communication with council?' ... But council went ... 'don't worry, it won't happen again.' And it will happen again.

Understandably, the group doubt that further incidents won't happen again, despite their regular communication with council staff. On the upside, one resident was so outraged by witnessing this undoing of the group's work that it inspired them to join the group to raise the profile of the issue and prevent further incidents from happening.

Managing community tensions

This chapter began by asking how the community responded to the removal of willows and resulting landscape changes. As you may have begun to suspect from the history of willows in Australia and Candelo more specifically, there are most certainly those in the community who grieve their absence.

> People who look back at how beautiful it was when it did have willows, but before there was a major explosion of them. Before they started to reproduce ... it was a beautiful English-looking creek. So, there are older people who still look back on that with nostalgia! And pity. They think it's a shame that it's now looking like this.

There was also a sense that the vegetation that replaced the willows, like reeds, made it no easier for people to access the river,[55] leaving some wondering whether the effort had been worthwhile.

These sentiments are a concern for members; they are keenly aware that not everyone in the town is supportive of their work. Comments on social media and in personal conversations are an ongoing challenge that the group faces in targeting a plant that has been so prominent in public discourse about Candelo for over a century.

> On the Candelo Connection [Facebook page], you'll get, 'Why don't you get the weeds out?' or 'What's the point of doing this, or that, or the other?' New people coming into town with new ideas: 'I remember how it was when the willows were there. What a shame it's like it is now.'

This tension was not only noticed by members of the group who had been present when the willows were removed. It was still being experienced by newer members, who joined many years after the willow removal had taken place.

> [I]t was quite challenging for me, actually, because she was just saying, 'A plant's a plant, why don't you just let it all sort itself out?' ... she had a pretty negative attitude to what we're doing. She sort of wondered why the willows had been taken out ... it was an uncomfortable, challenging conversation.

This conflict about whether willows are a weed or not and how the creek should or should not be managed, dating all the way back to the foundation of the group, has resulted in multiple efforts by Landcare to educate the broader community about the creek. This goes beyond the education efforts of other Landcare groups in this book, which mostly focuses on plant identification.

In Candelo, the group has invested significant time and resources in educational activities and materials. Beyond the signs installed and replaced just upstream of the traffic bridge, stalls have been held at market days and a photo muster invited residents to bring their old photographs of Candelo so that they could see what it looked like previously. This activity was specifically designed to address the 'resistance within some of the older locals'. More recently, the group set up a Facebook page to communicate the history of the creek, how it has changed over time, and how new vegetation, such as reeds, are a native species that help to filter water and provide habitat. Occasionally posts promote upcoming Landcare events, but mostly it is used as an educational opportunity.

> People who've been here a while say, 'Oh, what are you doing to the creek? It's turning into a mess,' or whatever. [The Candelo and District Landcare Facebook page] was an effort to try and just communicate more with people about the history of Candelo Creek, as a geomorphological entity, as a thing.

As part of this educational effort, the group has been proactive in engaging Candelo Public School students in educational activities around the creek. Beyond organising tree planting events with the students, activities over the last two decades have included the creation of educational materials to discuss the creek and how it changes over time.

Overall, members are careful to stress that they mostly receive positive comments that outweigh negative comments: 'People do love the creek, and they love the work, and they're thankful.'

This was confirmed in interviews with non-members who live within and beyond the community, who are positive about what the group has achieved.

> I think they've done pretty well. The water, there's a decent amount of water and it seems to flow quite well. You don't see too much of the invasive stuff, and they've done some good planting along the banks. Yeah, I think as individuals, they've got their ticket into heaven.

Philippa

A chapter on Candelo and District Landcare couldn't be complete without recognising the role that Philippa has played in leading the group over much of its duration. Every member I interviewed remarked on the invaluable contributions she made to the group: 'We've had one constant: Philippa. So, that has been incredibly helpful.'

Philippa joined the group soon after the public meeting in 2002 and took over the leadership from David in 2008. Since then, she has been working tirelessly to rehabilitate the creek, applying for multiple grants, liaising with the local council, South East Local Land Services and the Far South Coast Landcare Association, among many other organisations. She is not only passionate about the creek, and highly knowledgeable about plants but also has spent considerable time researching the history of the area.

> Number one, I think, is her extensive knowledge of native veg[etation]. She knows her stuff. Two, is that she has spent a lot of time trying to understand the changes to the creek and to Candelo with the last century and a half, so she's got a really good grasp of the local history. I think that's really important. And she has a talent for inclusiveness. She's able to get along with just about everybody.

Members remarked on the way Philippa deftly handles tension within the community about how weeds and the creek should be managed.

She is also someone who makes members feel at ease, regardless of whether they consider themselves to be social people. She helps people to feel included by making the group and its activities flexible enough for people to participate in whatever way suits them best: 'Part of the advantage of the Candelo Landcare is that you don't feel you're obligated to turn up.'

Others commented on Philippa's vision for the creek and how they appreciate the direction she sets, meaning they can simply come along and work.

> I'm actually really happy with how the group operates. I think what we have is a particular focus. We have a communal area, we've got set boundaries. We've got some flexibility around people coming and going. A lot of flexibility.

In 2023, Philippa's dedication to the group was recognised when she was awarded the Bega Valley Citizen of the Year.

Perhaps the highest praise for the work Philippa and the Landcare group has achieved comes from those who have been quietly observing their work over the years. In response to a question about whether the group has been successful, one interviewee replied:

> Well, just the change in Candelo Creek, for starters, with the removal of - there's a bit of a discussion about whether willows, for instance, are good in creeks or not, so the Landcare group decided to get rid of willows, which has created a beautiful flow back in the creek, and pools and stuff, which is wonderful. It really does work. From a really congested creek to one that has very good flow now, and beautiful big pools. I feel like it's really working well. So that's one example. And tree planting along the creek banks and things like that I think is wonderful. It provides an amenity for all, not just for people but for the wildlife. Birds. And it's also holding the banks together. And the general rule that people don't graze in creeks now, as well. I love that.

Continuing to care for Candelo Creek

The story of Candelo and District Landcare provides lessons for those who want to remove plants that are loved and loathed, often referred to as conflict species. Even after 20 years, there are still those in Candelo who lament the removal of willows. It isn't easy to change people's attachments to specific plants, the aesthetic they create or the other benefits they provide, so those who wish to remove plants that others consider beneficial need to be prepared for resistance, not just over the short term but over the long term. Agreement may never be reached on whether some plants are weeds or not, or whether their removal is beneficial.

In the case of Candelo, the group has been shaped by the ongoing conversations within the community about how best to manage the creek and what plants do and do not belong, their impacts and benefits. Working on highly visible public land has been challenging yet rewarding. For those who have spent years working to remove weeds, the benefits come from seeing others in the community engage differently with the creek. Overall, the group's weed removal work has afforded children and adults alike a new way of interacting with the creek, the lifeblood of the town.

9

Towamba Valley Landcare: a valley-wide effort to control serrated tussock

From afar you might not notice the tell-tale signs that suggest there is something special about the Towamba Valley. If you happened to drive through this remote valley on the day of their 6-monthly Serrated Tussock Working Group meetings, you might have thought there was a field day or other such event on. There would be a grassy paddock filled with utes and 20 to 30 people sitting in a circle in a big farm shed having a chat.

What happens here is quite remarkable.

I first came across the group at one of these meetings in March 2009. I was doing my PhD and had interviewed dozens of farmers and government staff in NSW and Victoria trying to understand why they did or did not work together to manage serrated tussock (*Nassella trichotoma*). Up until that point, the answer, time and time again, had been that it is uncomfortable to talk with others about serrated tussock.

Any mention of serrated tussock to neighbours may cause offence, taken as a criticism of how they manage their land. A government weeds officer might deem you to be recalcitrant in controlling serrated tussock, paving the way to fines and legal battles. The safest bet is to say nothing, or to only say something to those who you really trust. In short, serrated tussock was taboo.

So, you can only imagine my surprise when I joined the Towamba Valley Landcare circle to find that a group of farmers, weed officers, NSW National Parks and Wildlife Service rangers, and private and public forestry staff were openly talking about serrated tussock. And not only were they talking about serrated tussock, but instead of laying blame on those who weren't doing enough, they were sharing their concerns and challenges.

It seemed surreal that this Landcare group had managed to break free of the social norms that discourage farmers and government staff from talking to one another about weeds.

The group had somehow flipped the script on serrated tussock; instead of being viewed as an individual problem it was seen as a shared responsibility. As a person external to the group explained:

> How it started off was helping people with issues, mainly it was serrated tussock, and they would find the problems, come to our meetings – and the meetings were big, forty-something people, and it's the biggest Landcare I've ever been to ... it was helping people that couldn't do it.

The question that I pondered then and since, is how did they do it? Was it something special about the people in the Towamba Valley? What did it take to get a group like this going?

In 2009, they had been working together for ~17 years. Now, another 17 years on, this chapter not only reveals how they've done it, but what else the group has achieved in this time.

Serrated tussock

The history of serrated tussock is like that of many of other weeds; it was introduced to Australia by accident. Investigations suggest this pest plant was first introduced through equestrian equipment – saddle packing – that was imported from South America in the early 1900s.

The weed was first discovered around Yass, ~50 km north of Canberra, in 1935. At the time it was found by the noxious weeds inspector across three properties and estimated to cover 92 ha.[1]

Serrated tussock is particularly difficult to control because it is a grass. For those in the weed business, grasses are the most maligned because they are the hardest to spot and prevent spreading.

One of the challenges in identifying serrated tussock, is that it looks much like native Australian tussock grasses. It takes a trained eye to be able to spot it and notice its spread.

> Recognising it. That would have to be the single most difficult thing with serrated tussock. A lot of people don't know what it looks like. And it takes a bit of field experience to get to know what it looks like.[2]

Its spread is facilitated by mature plants that can produce up to 100,000 seeds each year,[3] which can be carried up to 20 km by wind.[4]

In NSW, serrated tussock spread from Yass across the central tablelands up to Armidale in the north of the state. It also spread south, along the Barton Highway, throughout the Australian Capital Territory (ACT) and down the Monaro Highway to Cooma and down the NSW coast towards Eden.

Those who have witnessed its spread speak to the way in which it thrives after droughts. The drought of 1937–1945 coincided with the rapid spread of serrated tussock in NSW and it being declared a 'statewide pest' soon thereafter.[5] More recently, attention was drawn to the spread of serrated tussock during the 2001–2009 Millennium Drought.[6]

By the mid-2000s, serrated tussock was documented to cover 2 million hectares of south-eastern Australia, costing landholders over $45 million each year in lost production and weed control costs.[7] When climate change predictions were taken into account, it was estimated that serrated tussock could eventually spread to cover 30 million hectares, an area of land comparable with the size of Italy.

Beyond the ease with which it spreads, there are several reasons why graziers and others are concerned about serrated tussock. For those who raise sheep and cattle, serrated tussock is known for having little nutritional value and the potential to poison livestock that graze on it.[8] Serrated tussock is also considered to be an environmental weed because it threatens the integrity of native grasslands and conservation areas.[9,10] Beyond its agricultural and environmental impacts, there are also safety concerns because of the fire risk it poses. Research on the impacts of serrated tussock suggest that it not only increases the intensity and potential spread of fires but also lengthens the fire season where large infestations occur.[11]

While there have been government programmes in NSW that encourage and support private landholders to control serrated tussock since the 1930s,[12] there are few places where serrated tussock has been prevented from taking over. Together, these features of serrated tussock made it a prime contender for the inaugural list of 20 Weeds of National Significance.

Control serrated tussock or else ...

For many years, local and state governments attempted to take a hard line on serrated tussock and on those who were not doing enough to control it. Like many weeds that came before it, serrated tussock was added to local and regional noxious weeds lists, meaning that landholders had a legal obligation to control serrated tussock and government staff had a right to inspect properties and evaluate the effectiveness with which serrated tussock was being controlled. Landholders who were found wanting could be issued with notices directing them to control serrated tussock, fined for not complying with directions notices and have contractors sent onto their properties to control weeds.

While this may seem like a case of government taking its powers too far, it was often farmers themselves who were demanding that governments take a hard line on those who weren't controlling serrated tussock. Many saw it as the role of government to bring the worst offenders into line. Indeed, weed officers frequently received phone calls from farmers telling them about neighbours who weren't doing enough to control serrated tussock, asking the weeds officer to come and inspect their neighbour.

Volunteers for community-government partnerships, such as the NSW and ACT Serrated Tussock Working Party, would often take up such roles because of frustrations about neighbours not doing enough to control serrated tussock.

What my research in 2008 and 2009 revealed was that this hard-line approach to serrated tussock had a deleterious effect on relationships between neighbours. Landholders were worried about talking to their neighbours about serrated tussock in case they took the conversation as a negative judgement about their ability to control weeds. Those who were new to an area were worried that if they asked for help identifying a plant that there was a risk they would be reported to the weed officer.

Some local governments in NSW tried to challenge the status quo, organising aerial spray programmes across multiple properties.[13,14] In these programmes, the local government would act as the go-between, finding landholders who were happy to join the programme and then coordinate days and times for the multi-property herbicide application to take place. But even then, challenges remained with what to do about properties who didn't want to join in the collaborative effort.

While the government is still looking for solutions and researchers have explored the potential for collective action to form,[15] there have been at least three examples of farming communities successfully collaborating to manage serrated tussock.[16] Two of these groups – Tussock Tamers and MacLaughlin River Landcare – were highly successful at bringing landholders together to control serrated tussock for 5 to 20 years, respectively. Sadly, both groups have folded. The third well known group, Towamba Valley Landcare, has been going for over 30 years, and is still going despite reaching a crisis point in 2022.

Way down south not very far off

Towamba has long been a place of cultural significance. It is found along the Bundian Way, an Aboriginal route – always was, always will be – that forms a shared pathway across Yuin, Ngarigo, Jaitmathang and Bidawal Countries, connecting Taragangal (Kosciuzko) and Bilgalera (Fisheries Beach).[17,18]

Towamba is almost equidistant from Sydney and Melbourne, ~550 km from each, located about half-an-hour inland from the coastal highway that links these two major cities and a similar distance from the NSW-Victoria state border.

The valley is nestled in between the escarpment of the Great Dividing Range and the Sapphire Coast. The rural landholdings are mostly used for cattle and sheep grazing and are surrounded by national parks and state forests including Mount Imlay (Balawan) National Park, Towamba State Forest, Nullica State Forest and Egan Peaks Nature Reserve.

The Towamba River and Towamba Road run through the valley. Towamba is the largest of the communities in the valley, with a population of ~250.[19] It has a township of sorts, with a sportsground, community hall, pay phone and public school. Other communities that are encompassed by the Landcare group are Burragate (~100 residents)[20] and Pericoe (~20 residents).[21] The number of students enrolled at the 162-year-old public school gives a sense of the size and composition of the community. It has consistently had ~20 students in recent years. Some students are the fourth generation to have attended the same school.[22] Towamba also has an active Progress Association and a multi-generational Rural Fire Service.

Like many of the groups in this book, Towamba Valley Landcare is in an out-of-the-way place. But one of the things that sets it apart from the other groups is that many of its members have been and continue to earn an income from farming. Indeed, over the last 20 years, beef and cattle farming have been the most common form of employment in the area.[23,24] While subdivisions have brought newcomers into the area in recent decades, the main motivation of this group has been to protect and enhance farming land; the focus is on using land sustainably. It is one of the few Landcare groups remaining in south-east NSW that focuses on managing agricultural weeds collectively.

Four farmers

Early accounts of Towamba Valley Landcare recall how on one day in 1992, four farmers were in a shearing shed and realised they had common concerns about the spread of serrated tussock, the threat it posed to their farms and the community more generally.[25] They had heard about serrated tussock from a state government agronomist and realised that they had better act to get on top of this weed threat that may have already been spreading into and across the valley for 40 years.[26]

At the time, the Landcare movement was beginning to take off across NSW. The four farmers, in discussion with the agronomist, saw Landcare as a useful medium for addressing their concerns about serrated tussock, as well as a range of other weeds and land management issues.

The Landcare facilitator, funded by what is now the Far South Coast Landcare Association, invited residents to come along to a meeting in the shearing sheds to find out what Landcare was about. There were 30 to 40 people at that meeting and soon thereafter the group had amassed 85 members.[27]

While the group was started because of weeds, there were other land management issues that also concerned them, such as pest animal management and water sharing. A document that holds the vision for Towamba Valley Landcare lists eight wide-ranging goals:

- To use land sustainably.
- To manage land in a way that is resistant to weed invasion.
- To prevent any spread of weeds to neighbouring properties.
- To eradicate any new invading weeds.
- To have no serrated tussock seeding.
- To reduce the incidence of African lovegrass.
- To manage pest animals effectively, especially rabbits.
- To share water resources according to our water-sharing plan.

It is unclear when this vision was developed, but it speaks to diverse concerns of group members, not only about the issues on their own properties but also minimising their impacts on others.

While weeds feature prominently in the list of goals, Towamba Valley Landcare has consistently recognised through its form and function that weeds are not the only issue of concern. Already in 1994, to recognise multiple goals, the group formed a Serrated Tussock Working Group that would focus specifically on serrated tussock management in the valley. While there was overlap in the membership of the Landcare group and the working group, their activities were run in parallel. The meeting I attended in 2009 was one of the 6-monthly meetings of the working group. More recently, the two groups have become one.

The driving forces

There is limited documentation available of working group activities from the 1990s. What does exist suggests that in the early days the group was mostly focused on supporting one another by raising awareness about serrated tussock. New residents to the neighbourhood would be door-knocked to be told about weeds and about the Landcare group.[28] As one member explained:

> The overall aim, initially, was to educate people who were new to the area about weeds in this area, and how to cope with them and deal with them.

This educative starting point is common in the weed management space. It is often assumed that people don't control weeds because they don't know about them, don't know how to identify them and are unaware of their legal responsibility to control them[29] (under the NSW *Biosecurity Act 2015* everyone has a general biosecurity obligation to prevent the introduction and spread of weeds). Thus, many groups begin by developing brochures and holding weed identification sessions.

At some point in the early 2000s, interest in Towamba Valley Landcare waned; membership numbers had dwindled to four.[30] People were frustrated by what they saw as a lack of commitment by various landholders and the local council to deal seriously with the serrated tussock threat. The classic collective action dilemma of what to do about the weakest links had come to the fore: what is the point of controlling weeds if your neighbour doesn't?

Burnt out and demoralised, the group looked as though it was about to fold. Then two things changed that would revitalise it.

The first was that one of the new weed officers for the local council lived nearby. He was a farmer who was familiar with serrated tussock from his own property and could appreciate the group's concerns about past weed officers not doing enough to encourage landholders to work on serrated tussock.

He began to work with the group, supporting them in their activities. He provided resources by way of herbicides, loaning equipment and putting on barbecues for community events. He would talk to all the landholders individually through his inspection programme and encourage them to join the group. Within the council, he had the support of his manager to take an approach that met the needs of the community. This support was critical.

An article in the local *Wyndham Whisper*[31] explained the positive view that the group had about council at the time.

> The council supports our actions and they are the responsible body. Hopefully neighbours are controlling their weeds or will make the effort if asked. Council can step in where recalcitrance rules.

The second important change was that the group was successful in obtaining funding from the Southern Rivers Catchment Management Authority to pay a respected local farmer part-time to oversee and deliver diverse land management projects. The farmer-coordinator sought and obtained funding for grants that helped individual landholders develop and implement weed management plans, supported weed management activities along roads and river corridors that were seen as vectors for the spread of weeds, and engaged Koori[32] work teams to undertake weed control work on private and public properties.

In 2014, an article published by *ABC News*[33] described the engagement of the Koori team as being 'a great part of the reconciliation process in our valley' and in 2017, an article in the local newsletter explained:

> In fact, our projects over 11 years have provided work for 39 different Aboriginal youth working for more than 1000 person days treating weeds. Often beside landholders and our TVLG [Towamba Valley Landcare Group] volunteers. Most of these

> people have gone on to full time employment ... We have had snake bites, heatstroke, hypothermia and lost workers but these young ambassadors for their people seemed always to take it with a smile and in turn all enjoyed getting on to Country. Many cultural artefacts were discovered and left where they were. They were all very respectful of their Ancestors who used to manage and care for the land and hope we can both respect that history and continue to care for and manage the Country well.[34]

With these two changes, Towamba Valley Landcare found a new rhythm. Each year the Serrated Tussock Working Group would organise two get-togethers. Everyone who managed land in the valley was invited. This not only included private landholders but also public land managers, including NSW National Parks and Wildlife Service, Forestry Corporation of NSW, private forestry and the local council. During this time, it was estimated that 60 per cent of the land in the valley was being managed by committed Landcare members.[35]

At these meetings everyone would take a turn explaining what they had done to contribute to the shared goal of having no serrated tussock seeding. The meetings would start with the local weed officer explaining what they had been doing to support the aim. This included an overview of the number of inspections of private property undertaken and the weed control activities that council had undertaken on public properties it was responsible for managing. Then everyone else would take their turn. They would explain the type of weed control work they had done – spot spraying, aerial spraying, chipping, hand weeding, improving pastures and tree planting – the types and concentrations of herbicides they had applied and whether it had been effective. They would speak about challenges they faced in controlling serrated tussock, such as having trouble identifying plants, rain events that either prevented herbicides being applied or that gave plants a boost. Others could then contribute their suggestions that might help the property owner in the coming year. This was social learning in action.

Before each person ended their turn, they would be asked by the long-term farmer leading the session to reflect on whether they had prevented serrated tussock from going to seed, as per the group's primary aim, and then they were thanked for the work that they had done.

Invariably, the group also discussed who wasn't present and the work that they had or hadn't been doing. They would talk about whether those not present had weed management plans in place. If there were concerns about places or people the weed officer had missed, they would be identified at the meeting.

These meetings would also be an opportunity for the group to reflect on anyone who was new to the area or those who were usually active but might need assistance, and how they could be supported. The group was very clear that they were only willing to help people who were committed to the cause of the group: 'the ones who the group really supported helping were ones who were trying quite hard themselves but weren't getting ahead, and mostly they were Landcare members.'

Beyond the meetings, the farmer-coordinator regularly shared information about weeds and other land management issues in the local newsletter, the *Wyndham Whisper*, as well as in news media across the region, such as *ABC News* and *Bega District News*. These articles provided photos of weeds, descriptions of various ways of controlling them (with and without herbicides) and advertised group activities. They would always include contact details for the farmer-coordinator, so that people could contact him with further questions or get access to weed control equipment.

This description provides some insights into how the working group came to be organised and how it was related to the broader group. Three important and interrelated questions remain. How did the group overcome the social norms that discourage rural landholders from talking to one another about weeds? How did the group extend its reach from weeds to other issues? And how did the group adapt and remain relevant over time?

Removing stigma

There were two ways the group removed the stigma of having weeds on their land. First, they focused on building relationships among landholders in the Towamba Valley. Members reported that 'number one is to try and make friends with them'.[36] The farmer-coordinator was known as 'Derek the doorknocker' because he was proactive in reaching out to landholders in the valley. But he wasn't the only one. Others in the group were encouraged to talk to new landholders and invite them to join Landcare. Members fondly recalled being approached to join the group by Derek, their neighbours or even the previous owners, and being grateful that someone came to talk to them about the weeds they had been blissfully unaware of.

To support relationship building, the group ensured meetings were accompanied by food. Daytime meetings led into barbecues and evening meetings ended with supper. Having food means people are more likely to join and it paves the way to friendship, 'it gets everyone together and gives us a chance to talk about things socially, which is good'.

The benefits that came from these relationships ranged from supporting one another in times of hardship to a sense of belonging. As one farmer explained:

> When I was farming, for example, I broke my shoulder. Those fellas came in, they shore the sheep, they castrated the lambs, they tailed them. They were incredible. So, I feel that I owe them something back.

But these relationships in and of themselves weren't enough to overcome the stigma of weeds. The group also explicitly acknowledged that weeds are a community-wide problem. Shifting the focus away from the individual creates an environment in which it is not only acceptable to own up to having weeds, but it is the socially inclusive thing to do.

> It's not the end of the world if you've got lovegrass, but it's important to put your hand up and say, 'Hey, I've got lovegrass

> here, and can someone help me fix it?' It's more of a crime to pretend it's not there ... it's a community challenge. It's not an individual challenge, and dealing with it together as a team is the only way in which we're likely to overcome it.

The valley-wide meetings were key to communicating and reinforcing that weeds are a collective and shared responsibility. By participating, landholders show their commitment to the communal goals of minimising and preventing the spread of weeds and that they are there to support one another.

> I think the effective thing was that by everyone meeting together in a group like that, it took away the stigma, because you recognise that everyone's got weeds and you're not just the main serrated tussock grower of the area, and we would make comments to each other about what had worked and what hadn't worked or try to help each other.

The meetings also demonstrated the progress being made towards the communal goal and provided reassurance that their efforts were worthwhile. Almost everyone I talked to agreed that the group has been successful in minimising the spread and impacts of serrated tussock: 'serrated tussock's still there, but not to the extent it was. That was a terrific effort to control it.'

Beyond the 6-monthly meetings, the group got together to control weeds along roadsides and river corridors. These roadside and river walks are fondly recalled by members because they provided an opportunity to get close to the 'beautiful pristine river' and 'wander along the roads on a beautiful sunny day'. The river walks presented an opportunity for children to join the group and to end with a barbecue. The walks provided a tangible opportunity to work together to prevent weeds spreading along these corridors and onto farmland.

> We'd meet once every six weeks or so, near Bluey's Bridge ... and we'd work right down to the [next] bridge, on the river. Mainly [African] lovegrass. We had up to 22 people and we had kids with us ... Saturday mornings ... We'd start at, say, eight

o'clock and at midday we'd have a barbecue ... It was great! ... That worked for some years. That was terrific.

An article in the local *Wyndham Whisper* in 2007 also conveys the joy of these roadside working bees.

> Twenty volunteers spent an idyllic Saturday morning taking the seeds off and spraying [African] lovegrass on the Rocky Hall Road and in the Towamba River. This completes our autumn program on all roads south of Mt Darragh Road.[37]

Beyond reflecting on group activities, articles in the *Wyndham Whisper* acknowledged the efforts of people in the broader community who were working to support the groups aims.

> A big THANK YOU to all of those who pulled fireweed so assiduously along our roads an on their properties. A special thank you goes to Genna and Aleisha from Wyndham Pony Club who have worked to keep public areas around Wyndham clear of fireweed, pulling weeds after school! This is a huge encouragement to those who are trying to keep their own land weed-free.[38]

Through such articles, the farmer-coordinator communicated expectations about the shared responsibility of weeds beyond group members. The benefits of such work were celebrated, further reinforcing the need for, and importance of collective weed control efforts.

> Our most important achievement, the attempt to reduce all the hotspots of [African] lovegrass and serrated tussock within the valley. Most of these sites now are at 10 per cent of original infestations but we will need another 10 years follow up.[39]

These efforts were clearly effective. In 2014, Serrated Tussock Working Group meetings continued to have at least 40 people present and 60 financial members who were committed to 'swap[ing] ideas and work[ing] together to tackle the problem'.[40]

Beyond weeds

It is evident that Towamba Valley Landcare has been highly effective in bringing the community together to collectively discuss and face many land management issues. This has included organising farm planning workshops, holding field days about 'different cropping methods, different animal management methods', fencing off rivers to exclude stock and allow tree planting and native regeneration, pest management, such as rabbit burrow ripping, fox baiting, feral pig trapping and developing a wild dog management plan.

Some of these activities have had benefits far beyond the valley. For example, Towamba Valley Landcare's efforts to control feral pig population in the mid-2010s – spanning 20,000 ha of private and public land – were showcased throughout the NSW South East Local Land Service region. The ways in which the group shared information and coordinated pest control activities were recognised as being 'critical factors in the success of the program so far'.[41]

Towamba Valley Landcare was also involved in the development of the wild dog management plan for the area spanning the Bega River to the Victorian border and up the escarpment past Bombala. Their ethic of engaging diverse stakeholders was embedded in the plan. They recognised the different ways that 'wild dogs' are defined, including recognising their cultural significance to Aboriginal groups.[42]

While these activities are appreciated by members and show the potential for collective weed management principles to expand to other issues, others are concerned that focusing on too many issues undermines the close social fabric of the group and impact its momentum.

> Every time we tried to sort of go broader environmental, it tended to lose some of the hearts and minds ... in the old days, I think it was called holistic management, now regenerative agriculture. If you start talking too much about those concepts you tend to lose the hearts and minds that hold it together.

Conversely, when I talked to someone who wasn't a member of the group, they pointed to its weed focus as a reason for not joining. Hearing that the group was now hosting discussions about carbon sequestration meant they were more inclined to join future meetings. This shows the tension between finding topics that a critical number of people are interested in and making sure that the topics selected meet the needs of those who are most active and passionate about keeping the group going.

Adapting to changes

Just as the combination of a new weeds officer and funding breathed new life into the group in the early 2000s, a new resident has reinvigorated the group since 2022.

Like most of the groups in this book, Towamba Valley Landcare was affected by the 2019–2020 Black Summer bushfires – the Border Fire swept through the area in early January 2020 – and the COVID-19 pandemic. Yet it was already facing challenges before these unprecedented events.

In 2015, when a colleague and I interviewed members of the group, some were already worried about its future. One of the challenges identified then was continuing to find funding to pay the farmer-coordinator and for weed control activities. An article in the *Wyndham Whisper* in 2017[43] recognised that '[w]e unfortunately have no one who wants to run the treadmill of applying for and managing funds and the Landcare funding is a bit thin at the moment anyway'.

Added to the challenge of securing funding was the introduction of the NSW *Biosecurity Act 2015*. This Act fundamentally changed the ways local governments dealt with widespread weeds. Weeds like serrated tussock, African lovegrass and fireweed (*Senecio madagascariensis*) were no longer listed as priority weeds and weed officers moved away from enforcing compliance.

Around the same time, the local weeds officer retired. While the person who replaced him was keen to support the group, they were not based nearby and were less involved.

The changes to the Biosecurity Act and staff turnover have meant that some members feel that the group's relationship with the local council is no longer as supportive and reciprocal as it once was. They feel they have been let down because the council no longer attends meetings or supports their activities, although other members feel that they don't need the council to run effectively.

Finally, during this period the founding members of the group moved away and the farmer-coordinator felt it was time to pass on the leadership. The loss of this drive and momentum was noticeable.

An extraordinary meeting

The night I revisited the group in 2022 it was eerie driving along the road from Eden to Towamba. It was a cold and crisp June evening. There was a red hue to the sky and the headlights brought into sharp relief the impact of the Black Summer bushfires on the heavily wooded areas of the Towamba Valley. The black tree stumps stood tall, lined up along the road. The fire had clearly burned hot. There were few leaves emerging from the char. The almost full moon shined brightly on the trunks of far-off ridges, making it look as if the mountains were covered in short hairs standing on end.

In the community hall, around 20 residents drifted in, unsure about the future of their group. Some of the faces were familiar, some even remembered me and my PhD research from 13 years earlier. It was bittersweet to see them again.

With the residents came plates of home-made food, so much food that there was barely a place to put it all. Sweets, sandwiches, tea, coffee. No-one touched it though. The topic of the night was a serious one. The mood wasn't sombre, but respectful.

Derek kicked off the meeting. He brought the gathering up to speed on the projects that had been finished recently, dealing with weeds across 52 properties, the financial costs of continuing to run the group, the lack of group momentum. The question for the group that night was: should they keep going?

Derek thought that only five or six people would come, but with 20 or more sitting around the table there was already an inkling that this group was valued by the local community.

One by one, each member spoke about what the group meant to them, acknowledging the challenges and the drain it is for those in leadership roles, but that the group has been 'a really good thing for the valley'.

Members spoke about how the group helped them when they first moved to the valley, to learn about what the land management issues were, and to meet the others in the community. One after the other, members confirmed that they believed that 'Landcare is really important to this community' and that they needed to keep it going. They were concerned that if the group dissolved, then people would go back to working on their own, isolated and fragmented.

They reflected on the various grants they had been awarded and the progress they had made in controlling serrated tussock and keeping other weeds at bay, the way they had started small and then expanded the area that they worked on. They also reminisced about the river and weed walks. Unlike other groups, who are unsure about whether they had made progress on weeds, this group was unequivocal. They had reduced the spread and density of serrated tussock.

They also spoke about their appreciation for Derek for coordinating the group for as long as he had, for his work in supporting the Koori teams, and recruiting new members.

It took more than an hour for each person to have a chance to tell their Landcare story. There was so much feeling, passion and appreciation conveyed that there was no way the meeting could end without someone stepping forward to lead the group. Soon someone, a newer resident, offered to continue the group, to apply for grants and to organise the next set of activities. Others put up their hands to join the committee and take on a new community liaison position.

Once that was settled, it was time to tuck into the food that until then had been left waiting, untouched at the side of the room. The conversation ensured that the food was celebratory.

As I drove away that night, I cried. Tears for all the feelings that people had shared and for the sense of relief at knowing that such a marvellous group would continue.

Managing land sustainably

Since 2022, the group has reconnected with its overarching aim of managing land sustainably. It obtained funding for a Smart Farms

Towamba project and has run a series of workshops on agroforestry, all held at the community hall, each attended by 20 to 30 people with some coming from almost 100 km away.

In 2023, a year on from the extraordinary meeting, I attended a workshop on carbon sequestration at the Wyndham Hall. Lachlan Ingram, a well-respected local scientist, was there to give a presentation. Many in the community had been asking about the topic.

Testament to the ongoing relevance of the Landcare group, there were 30 people who attended this 1.5-hour presentation on a Friday night. Lachlan spoke about how emissions from grazing are calculated and ways farmers can calculate baseline emissions. He spoke about options for reducing such emissions but raised questions about unintended consequences of such efforts. Members asked questions as the presentation progressed. There was discussion about the ways weeds might affect emissions.

As you might expect by now, the meeting was well catered. A local caterer had provided an abundance of Finnish fish soup, roast beef casserole with potatoes, warmed bread, wine and rice pudding. After the meeting, everyone stayed and joined a long table to share dinner and catch up. Despite being far from home, I didn't feel like an outsider.

Last group standing

While Towamba Valley Landcare has had its ups and downs, it is a remarkable achievement that it has persisted for 30 years in the face of changing institutional support, the Black Summer bushfires, the COVID-19 pandemic and the loss of the founding members.

While it is likely that there are a confluence of reasons that explain its persistence, understanding the reasons why other groups folded may provide some insights.

Interviews with members of the now lapsed Tussock Tamers and MacLaughlin River Landcare suggest there is something unique about Towamba Valley Landcare.

First, obtaining funding to pay a local farmer to coordinate group activities is important because it recognises the time commitment required to run a group like this and provides longevity in a way that government agencies cannot. MacLaughlin River Landcare was reliant on a coordinator who was employed as a Community Support Officer with the Southern Rivers Catchment Management Authority. When the Catchment Management Authority was disbanded and replaced by South East Local Land Services and the funding ran out, the group had no-one who could continue to apply for funding and keep coordinating cross-property activities. Tussock Tamers also worked closely with the local council and South East Local Land Services. They worked together to produce promotional material. When their initial funding ran out, the local council continued the programme for several years. So the employment of a local person provided Towamba Valley Landcare with extra longevity and an ability to persist, despite institutional changes. It has enabled them to adapt to changing government and community priorities.

Second, there is something to be said for the way in which Towamba Valley Landcare brings members together regularly to talk about their common concerns and strategies for dealing with serrated tussock. Tussock Tamers adopted a different strategy; they encouraged active members to display signs on farm gates to indicate that they had completed a property plan. Members described these signs as a way of raising awareness of group activities, acting as a visual indicator of the extent of participation in the project.[44] To some degree, this was effective as events run by Tussock Tamers were well attended. Records show that 180 farming families participated in workshops and over 300 attended field days. Yet the scale of the membership base and the lack of regular meetings among the larger group may be what ultimately meant it was too big to be sustained over the long term.

Members of MacLaughlin River Landcare also mention group size as being one of the factors that may have undermined the longevity of their group. With too many people a group becomes 'unwieldy' and 'disorganised', it becomes 'harder to get that commitment'.

Beyond group size, reflections on MacLaughlin River Landcare highlight how collective goals, funding and evaluation can inadvertently create inequalities among members. The aim of MacLaughlin River

Landcare was to 'get rid of the big motherlodes of tussock'. Implicit in this aim is that some people have worse serrated tussock infestations than others. This was also made explicit in the way funding was divided. Unlike Towamba Valley Landcare, where funding was more likely to be allocated to members who were committed to contributing to the group goal, funding in MacLaughlin River Landcare was preferentially allocated to those with the worst infestations. This approach differentiated group members by how much serrated tussock they had. Over time, these differences became significant. Members became resentful of those who received funding but who did not meet the group's expectations about how that funding was used or how effective it would be.

> That's a pretty problematic way of divvying up the funding. We didn't say, 'Right-o. Everyone gets an equal share.' ... no-one complained, but I think underneath it there was a bit of, bloody hell, he's a big one [landholder]. He can afford to do it [weed control] anyway, but he's still getting funding.

This quotation reveals how differentiating between group members can undermine the social fabric of a group. When the founding members left the MacLaughlin River area, there were limited routines, processes and sense of collective purpose left to facilitate the upkeep of the programme by those who remained.

A lifeline and long-lasting legacy

No matter what direction the new leadership takes, Towamba Valley Landcare has much to be proud of; beyond its collaborative efforts to control many weeds and pest animals across private and public land, it has given people in this small and remote community a reason to come together. The value of working towards a common cause, of sharing food, caring for Country and one another cannot be underestimated. Indeed, it is this strong social fabric that keeps members committed to seeing the group succeed.

Yet there is something to be said about keeping a focus on weeds. For this group, controlling weeds is a shared experience; it connects everyone in the valley regardless of their reasons for owning a property or how they manage it. The act of controlling weeds shows a visible commitment to the broader community. There is also a temporal imperative each year to control weeds before they seed, which provides a strong reason to come together on this topic and regularly share information.

Working towards reducing carbon emissions, if this were to become a new aim for the group, also has broader benefits. But to make this the foundation of the group, is likely to require annual meetings at least for members to explain how they're working towards reducing their emissions, to make this effort visible and to continue to demonstrate solidarity. It may also be worth considering an activity the group can regularly do together to make progress towards this goal.

INTERLUDE: BEYOND VOLUNTEER ENTHUSIASTS

The stories in this book paint a picture of people volunteering for the greater good and without financial gain over the long term. Yet being volunteers, and environmental volunteers in particular, makes Landcare members distinct for three reasons.

First, Landcare volunteers are part of a minority of Australians who volunteer. It is estimated that between 14.1[1] and 26.7[2] per cent of Australians aged 15 and older volunteer with formal organisations or groups. The Australian figures are similar to those reported overseas. For example, in the United Kingdom, it is estimated that 27 per cent of people over 16 are volunteers, with 16 per cent of the population volunteering monthly.[3]

While the number of people volunteering for the Landcare groups in this book have largely remained stable, volunteer rates elsewhere are declining. Census data[4,5] and other social surveys[6] show a 4 to 8 per cent decline in the proportion of people who volunteer over the last 20 years.

Second, many Landcare members are remarkable because they are what I would term volunteer enthusiasts. Many members not only volunteer to control weeds as part of Landcare but also volunteer for a range of other organisations. Members talked about their contributions to the NSW Rural Fire Service, arts societies, sports clubs or Aboriginal organisations. They also volunteer for other local environmental organisations like BirdLife Shoalhaven, the Kangaroo Valley Environment Group and Sustainable Agriculture and Gardening Eurobodalla (SAGE).

This pattern is consistent with other research I have undertaken in coastal Australia. Coastal communities often have a group of 'community-minded' residents who volunteer for multiple organisations and have distinct interests, activities and relationships with the environment compared to other residents.[7,8] These volunteer enthusiasts tend to be more social, have more friends and feel a stronger sense of belonging to the community. This tendency for volunteering to self-perpetuate is consistent with broader social research, which finds that past volunteering is a strong predictor for future volunteering.[9]

Finally, Landcare members are noteworthy because environmental volunteers are much smaller in number than interest would suggest. A 2022 survey by researchers at the Australian National University (ANU) revealed that 23 per cent of people are interested in volunteering for an environmental organisation and yet fewer than 2 per cent of the Australian population actually end up acting on this.[10]

While most of the Landcare groups in this book discussed the challenges of recruiting new members, the data suggest that there is latent interest in their cause among the broader population. This begs the question: why don't people join Landcare?

Why people don't volunteer

Before I dive into the reasons why people don't join Landcare, let's take a brief look at what the academic research has found to be the three key reasons why people don't volunteer.

The first is that people don't have time, skills or physical capacity to volunteer.[11] In a recent ANU survey about volunteering in Australia, work/family commitments (41 per cent) and health reasons (18 per cent) explain why many people don't volunteer.[12]

The second reason is that people don't want to volunteer. This may be because it doesn't align with their values, they can't see any benefits of volunteering or don't want the social burden or stress that comes with doing so.[13] In the ANU survey, 'not interested' (22 per cent) was the second most common reason people provided for not volunteering.[14]

Finally, people don't volunteer because they haven't been asked.[15] The ANU survey suggests that 14 per cent of people don't volunteer for this reason.[16]

These three reasons were evident in interviews that I conducted with non-members across south-east NSW. Their answers provide some insights into what groups might be able to do to encourage them to sign up and participate.

I don't have time

Interviews with non-members indicate that one key reason for not joining Landcare is because of a lack of time. This may be because work commitments make it difficult or impossible to go to working bees or being a part-time resident means that participation is only possible on occasional weekends or being a member of other volunteer organisations takes priority. When I asked one interviewee whether they would consider joining Landcare, they explained: 'Not really, simply because my time's caught up with the voluntary work out there, and what I've got to do here, and I feel like that's enough.'

This response and those of other non-members suggest that conflicting priorities are the underlying reason why people say that they don't have time for Landcare. One candid response by a non-member to my query about what prevents them from joining Landcare shows how 'not having time' reflects personal priorities.

> Only the time factor, but you know, it's all subjective. I mean, time is such a subjective thing. Most people are saying they haven't got time, but it's all a matter of priorities, isn't it?

This quotation may explain why changing the day or time of working bees may have little effect on membership numbers. Some groups in this book, like Deua Rivercare, have considered changing the day and time of their meetings to increase membership. Other groups, like Upper Kangaroo Valley Landcare (Chapter 8), experimented with shifting working bees from a weekday to a weekend and found that this had no impact on numbers.

Recognising that the day and time of meetings may not improve membership might help Landcare groups refocus their recruitment efforts. There may be other things they can do to make Landcare a more appealing choice and bump it up the list of priorities.

I don't want to volunteer

Interviews suggested there were three reasons why people don't want to volunteer: methods of weed control don't align with their values, weed control is too physically demanding or they want to avoid social tensions.

Weed control should be done differently

One value-laden reason that non-members provided for not joining Landcare pertained to the use of herbicides. One non-member didn't join Landcare because they felt that they didn't use enough herbicides:

> I've never belonged to a Landcare group. Not because I'm against it or anything. Just haven't had time, or just the different outlook, different view on things. I think they don't like to use chemicals as much.

Mostly though, use of herbicides deterred membership. Two non-members explained that they have considered joining Landcare but have decided not to because they are concerned about the use of herbicides.

> Decades ago I was very keen to join the Landcare group, was delighted there was one in my neck of the woods. But when I came and people were using Roundup, and that's what they always did, I was privately appalled and didn't even stay for the first work bee.

> I disagree with them using chemicals in the waterways and stuff, even though they claim that they're water-friendly, aquatic-friendly and stuff. I think there's other methods that can be used.

This concern was common among interviews with non-members, even if they didn't directly attribute this as a reason for not joining Landcare. Indeed, among the broader population there is a growing aversion to the use of herbicides, with growing calls to ban common herbicides like glyphosate and paraquat.[17]

Some of the groups, like Tathra Land Care (Chapter 7) and Candelo and District Landcare (Chapter 8), no longer use herbicides now that they have effectively removed bitou bush and willows from their focus areas. However, contractors and council staff may still employ herbicides on these and other weeds on behalf of the group. Other groups, like Brogers Creek Landcare (Chapter 3) and Upper Kangaroo Valley Landcare (Chapter 4), have been explicit that they are happy for members to opt out of the use of herbicides. Making clear the reasons for using herbicides and the care taken in their use may be needed to further alleviate non-members reservations about joining Landcare.

Weed control is too physically demanding

One thing that Landcare offers that many other volunteer organisations don't is an opportunity to be outdoors. While this may attract some people, it also brings challenges.

Perceptions about the amount of physical work required to participate in working bees is one consideration that deters people from joining Landcare. As one person explained: 'I'm getting older. I think it's a younger person's thing.' Another person who had previously volunteered for Landcare but no longer does explained that an acute illness prevented them from volunteering.

> As soon as I came back in the valley [people asked], 'Oh, will you join this? Will you do that? Will you do it?' and I just had to learn to say no, and I couldn't, and the last three years I've had rheumatoid arthritis, so I'm just getting on the mend from that, able to move, and got a bit more energy, so I've really had to just measure where I'm putting my focus.

Most of the Landcare groups in this book have made a point of welcoming community members to be part of Landcare even if they can't wield a chainsaw or go trudging through creek banks and up hillsides looking for weeds. Some groups have had members who are well into their nineties come along to working bees, even if it is simply to wander along and be close to those at work or join in for the tea and cakes afterwards. Others have taken on roles in social media and communications.

This quotation from a Landcare member explains the importance of being inclusive:

> [A successful community weed management group is one that is] inclusive and inviting, welcoming, safe. That's a big thing. People come and they go, 'Oh my god, are we walking into that?' I'm like, 'Yeah. Well, you don't have to. There are other ways for you to participate if you can't get down on your knees or you've got repetitive stress, you can't use secateurs or whatever, you could maybe do the newsletter or you could help

> with the morning tea or you could set up the maps or the contacts with the landholders.' So I think having a variety of roles, because it's not everyone's idea of fun to get scratched up and covered in ticks and leeches.

Avoiding people

Some non-members indicated that they don't join Landcare because they prefer to work alone.

> I don't sort of associate with any groups, because I just do my own thing. I just identify things I think are a problem, and then just work away at them.

Others provided examples of specific people who they knew were members of Landcare and whom they wished to avoid. Clearly, there are limits to how much Landcare membership can grow given individual social mores.

There is no group in my area

There was no evidence in the interviews I undertook that people didn't join because they hadn't been asked. Instead, non-members discussed the challenges of finding out about Landcare opportunities.

Three non-members I spoke to had heard about Landcare and were keen to join but there were no groups nearby for them to join. Another non-member had considered launching their own Landcare group but didn't know where to start or whom to approach. Similarly, other non-members contacted the local council to find out about Landcare because they didn't know about the Landcare networks that span the region they live in. At the other end of the spectrum, one person I talked to believed that because the Landcare group was so successful that they didn't need any more members.

It seems that not being asked may play less of a role in determining Landcare membership than knowing where to start, who to talk to or believing that the group could benefit from extra hands.

Growing the Landcare movement

In 2023, Landcare NSW received an influx of funding from the state government.[18] This has resulted in an uptick in the number of posts on social media and Landcare websites that promote the work of local groups. This is important but unlikely to be enough to increase member numbers.

I would argue that there is work to be done in raising the profile of Landcare groups and networks beyond social media. Having perused the websites of many Landcare groups, I can confirm that many have scant or outdated information. Often there is only a contact name and number, with no details about when working bees are held or who can attend. There is considerable scope to make Landcare groups, their activities and successes more visible online as well as making it easy for potential members to find someone to call or chat to about joining or creating a group. Providing such information may make it easier for those who are keen to be an environmental volunteer to consider their closest Landcare group as a place to start.

There is also likely to be more that could be done to understand how Landcare membership is driven by local social norms. Most of the groups in this book are in communities with volunteer rates that are much higher than the state and national averages. Groups like Tathra, Deua and Upper Kangaroo Valley have volunteer rates that are more than double the state and national averages, while Brogers Creek is quadruple. In part, these figures may be shaped by the Landcare groups themselves, but they may also speak to a latent interest in these communities towards volunteering, and environmental volunteering in particular. Unpacking this relationship may reveal where there is untapped interest and where new groups may thrive.

Regardless of the extent of volunteering rates in communities or the amount of information available about Landcare online, what the experiences of the groups in this book show is the importance of regular and proactive recruitment. Having personal conversations helps to make residents feel welcome and address any concerns they may have about joining.

Recipe: Nutty Cake

The social side of weeding can be just as rewarding as the satisfaction of seeing a weed free landscape. Cakes like these are core to building social relationships.

Ingredients

240 g Brazil nuts
165 g pecans
185 g pitted dried dates
90 g dried figs
85 g raisins
30 g currants
30 g mixed candied citrus peel
100 g plain flour
½ tsp baking powder
165 g caster sugar
½ tsp salt
3 beaten eggs
1 tsp vanilla essence

Method

1. Line a loaf tin with baking paper.
2. Preheat oven to 120°C.
3. Mix dried fruits and nuts together in a large bowl.
4. Sift flour, baking powder, sugar and salt.
5. Add flour mixture to the dried fruit and nuts and mix well until the fruit and nuts are completely coated.
6. Add in the eggs and vanilla essence and stir until the fruit and nuts batter is sticky.
7. Spoon mixture into the loaf tin and press down firmly with a spoon.
8. Bake in oven for 2 to 2½ hours.
9. Remove from oven and leave in the tin for 10 minutes, then turn onto a wire rack to cool.

This is based on one of my Mum's favourite recipes, which I baked and shared with Towamba Valley Landcare on my most recent visit. My Mum called it a fruit cake, but it is nothing like the Christmas or wedding fruit cakes common in Australia. I have called it Nutty Cake because the whole nuts in the cake are its defining feature. Essentially, this cake can be made with many different variations of dried fruits and nuts that together weigh 825 g. My mum's version included 200 g glacé cherries and 130 g walnuts. I have tried it with dried apricots instead of figs, which also works well.

10

The importance of weeding together

When I started this research project there was little robust evidence to confirm whether collective approaches to weed control are more effective than individual efforts. While policies across Australia have been shifting towards more collaborative approaches,[1] there was only fragmented evidence of the benefits they can provide. What the stories in this book show is that long-term collective weed management provides significant environmental and agricultural benefits. Without such collective efforts weeds would have continued to thrive unabated, transforming landscapes along the way.

While some highly passionate and dedicated individuals might have persevered on their own, it is unlikely that individuals would have achieved the extent of changes delivered by the groups over the long term. The collective effort means people can come and go and help in their own ways, each building on the contributions of the people before them and knowing that others will join later to continue their work.

Overall, group members agreed that being part of Landcare was important because they could see the difference their group was having. Through Landcare they cared for the local environment. They removed weeds, flood-swept rubbish and bushfire-ravaged debris. They planted, watered and tended to trees, shrubs and grasses. They enjoyed working in creek beds and along rivers on sunny days. Through their purposeful endeavour they observed a positive change in their surroundings.

What has made such efforts effective over decades is commitment to nurturing community. My research sought to understand the way social relationships shape collective weed management groups over the long term. While the benefits of Landcare for social capital[2,3,4] and the development of social norms that welcome newcomers[5] have long been explored, what my research has highlighted are the intangible social benefits that come from being involved in groups like these. The food shared, relationships established and emotional support provided are not an afterthought, a happy coincidence or a resource to be exploited. They are the heart and soul of Landcare.

The group members consistently mentioned how much they valued the sense of connection. Landcare gives them the opportunity to meet people with similar and different aspirations and viewpoints. They can point to the friends they've made and how much they have learned. What is clear from the interviews is that the social, environmental and agricultural benefits go hand in hand.

The environmental and agricultural benefits of Landcare

Landcare was formally launched in 1986 so now is a good time to take stock of the expansive impacts this movement has had. While there has no doubt been many short-term successes, changes at a landscape scale take decades to achieve and maintain. As anyone involved in controlling weeds will tell you, getting on top of weeds is not something that can be achieved quickly or with one-off spraying of herbicide. It takes years to exhaust the seed bank and make sure that any new seedlings don't have a chance to set seed and create tomorrow's problem.

The chapters in this book have painted a picture of just how different the vegetation was in each of these seven communities in the 1990s. Back then the dunes in Tathra were covered in bitou bush (*Chrysanthemoides monilifera* ssp. *rotundata*). The creek beds of Candelo were awash with willows (*Salix fragilis* var. *fragilis* and *Salix babylonica*). Serrated tussock (*Nassella trichotoma*) was spreading across the Towamba Valley. Lantana (*Lantana camara*) was smothering the hillsides of Budgong, Tapitallee and Bengalee. And privet (*Ligustrum sinense*) was escaping the hedges and gardens of the Deua, Wattamolla and Upper Kangaroo River valleys, making its way along the creeks, through the rainforests and into national parks.

Brought in intentionally and unintentionally by people, these plants were changing landscapes. Most had been slowly spreading for years before a drought, flood or fire caused an explosion in their numbers. En masse they outcompeted native vegetation, made dunes more erodible, slowed the flow of rivers, changed the habitat available to birds and animals, and made farming less viable.

In many places, these weeds and others, continue to have such effects, but what the stories in this book show is that a small number of people, even just 10 to 20 people, who consistently come back to the same place can make an appreciable difference to their local environment. The progress might be slow, slower than we might like, but change is possible.

There are seven good-news stories in this book. Not only have the groups transformed landscapes but they have helped their communities weather severe climatic events better as a result.

Today, the dunes in Tathra are covered in coastal beard heath (*Leucopogon parviflorus*), banksias and wattles. The Landcare volunteers are now finding fewer than 100 bitou bush each year. Many are seedlings, occasionally a mature plant is found. This is a remarkable success story. No doubt, the envy of many other Landcare, Bushcare and Dunecare groups who continue to battle bitou bush along the mid to north coast of NSW. While it is unclear how bitou bush contributes to fires, the way it has thrived after coastal fires elsewhere[6] suggests that the work of the Landcare group probably spared Tathra from a massive explosion of bitou bush after the 2018 bushfire.

In the foothills behind Nowra, lantana is largely controlled on the properties of active Landcare members. Many species of birds are thriving and the trees that were planted by the group in 2019 are now over 2 m tall. While the reduction of lantana from properties in the area did not save members from the 2019–2020 Black Summer bushfires, they are likely to have made the fires less intense. Research in other parts of Australia has found that lantana increases the intensity of fires because it creates a more continuous layer of understory.[7] It doesn't bear thinking how much worse the Black Summer bushfires could have been had the group not invested so much time in clearing lantana.

At the most recent meeting of Towamba Valley Landcare, most members reported that they felt serrated tussock was not much of a problem and that they had it largely under control. They spoke of seeing a sprinkling of serrated tussock plants. Keeping serrated tussock in check has enabled other grasses to thrive during the wet years that followed the Black Summer bushfires, making sheep and cattle grazing enterprises more viable.

Candelo Creek is largely willow free, making it easier for the residents to access the water there. The river now probably flows in a way that is similar to the way it did before the introduction of willows. Other plant species have established and the way the river responds to floods has likely changed as well.

Even the much-maligned privet is less widespread in the Deua, Wattamolla and Upper Kangaroo River valleys. At the most recent Upper Kangaroo Valley Landcare Annual General Meeting, members noted that privet was fading into insignificance. Like the removal of willows from Candelo Creek, the removal of privet and other weeds from rainforest and riparian areas has enhanced their ecological integrity.

Changing vegetation and susceptibility to fires and floods are only part of the story of how Landcare builds community resilience to natural disasters. The social benefits are just as important.

The social benefits of Landcare

There was widespread agreement among the group members interviewed that social connection is what they value most about being part of Landcare. Far from being an organisation that only delivers environmental and agricultural change, Landcare provides an opportunity for deep and extensive social networks to form. These social relationships are as important in a time of increasing climate-related disasters as the on-ground work that helps to mitigate and respond to environmental changes. Indeed, these networks are remarkable not only for the support they provide before, during and after disasters, but also because they go against broader trends of declining volunteerism[8] and increasing social isolation.[9]

Landcare provides a forum for building foundational social relationships. By meeting regularly and working together to address environmental and agricultural challenges, there were two key social benefits that Landcare members valued: new knowledge and social cohesion.

Landcare members consistently reflected on how much they valued being able to learn from knowledgeable people. Tellingly, the one thing all the groups have in common is a member with expertise in plant identification. These people are incredibly knowledgeable about weeds and how to identify them, as well as having expertise in native plants and ecology more broadly. The people with such knowledge have often completed Technical and Further Education (TAFE) qualifications in horticulture or conservation, have run nurseries or have university-level ecological or botanical training. Some are long-term farmers or were local government staff who have learned how to identify plants over time.

What members of the Landcare groups valued was having an opportunity to learn how to identify individual plants, whether plants are endemic or not and how different plants interact, as well as improving their understanding of the ecological significance of their environment. Members learned about indicator species, soils and habitats. They learned about native animals, birds and reptiles. They learned about the history of their local area and how the landscape changes in response to environmental upheavals, such as bushfires.

Beyond learning about the environment, members also learned about the role that they can play in managing weeds and rehabilitating ecosystems. They learned management techniques with and without herbicides, from the experiences of others and through their own trial and error. Through learning by doing, they became more proficient in managing pest animals, tree planting, carbon sequestration, regenerative farming and agroforestry as well as other land management practices.

What is particularly useful about the knowledge shared through Landcare working bees, weed walks, plant identification days, and regular meetings is that it is highly place-specific and dynamic. By learning together, Landcare members amass a wealth of knowledge that can be readily shared and updated as the environment changes. They know which are the latest weeds arriving in a given environment and can work together to find the most effective ways of dealing with these new plants. They can draw on expertise from across district and regional

Landcare networks to determine where effort is most likely to be worthwhile and what can be left alone.

The opportunity to learn is a drawcard of Landcare. It is something that brings people into the movement and encourages them to stay.

The other key value of Landcare is the opportunity to meet like-minded people. The members I interviewed mentioned how much they valued the opportunity to get to know their neighbours and have friendships bloom. Such relationships evolve over time as people work alongside one another in removing weeds and planting trees, united by a shared concern for changes to the environment and a commitment to doing something about it.

For most groups, the opportunity to socialise over a cup of tea or coffee and a piece of cake was a major drawcard. Making time for cake provides an opportunity to reflect on work done and work that remains. It provides a sense of companionship, removing a sense of isolation that can come from living in rural areas, and nourishes (pun intended) an important support network.

Members talked about the goodwill and community spirit that comes from Landcare. There were countless stories of members coming together to help one another. Members supported landholders whose properties were burnt during the bushfires. They purchased food and supplies for those trapped by landslides and helped with shearing and lambing when farmers were unwell or injured. They also shared extra resources – wood for fires, excess fruit and vegetables grown, and biological controls for weeds. And then there was the emotional support. They provided a shoulder to cry on, listened attentively to one another, and consoled one another through grief, feelings of being overwhelmed, and fear.

Together they filled a gap in government services and overcame patchy or unreliable communication. They shared messages about road closures, fallen trees and warnings about rains, floods and fires. They provided early information about whether homes had survived bushfires and offered shelter to one another. They provided a place where those who are elderly and live alone could find connection and an opportunity to be active.

By working together in these ways, on weeds and more, the Landcare groups create their own social norms and effectively reduce the stigma that can come from having weeds on one's property. They create a safe space to be vulnerable, admit that weeds are an ongoing challenge and ask for help. It is for this reason that the social benefits of Landcare can't be separated from the environmental and agricultural benefits.

Time and again, I've come across places where people don't work together on weeds. They can't see the point. Often, they believe that there is likely to be little chance that weeds can be effectively managed, even with collective action, or that the reduction in weeds is unlikely to be worth their time. Such perspectives overlook the significant social benefits of getting together and jointly tackling a shared problem.

As this book has highlighted, beyond friendships, working together on weeds brings numerous shared benefits with regards to knowledge learned about weeds, other invasive species, environmental changes and the latest innovation in land management.

Overall, Landcare offers continuity in a way that many government departments can't because of changing political priorities, budgets and staff. The flexible nature of Landcare, able to be tailored by and to the people in specific places, is important. This movement is responsive to local circumstances and enables community resilience in the face of significant environmental change.

Ways of working together

Beyond demonstrating the benefits that come from working together on weeds, the stories in this book provide insights into the ways in which people can effectively work together over the long term.

While the groups may differ in terms of how they operate and how regularly they meet, there are many commonalities. Here I identify seven attributes that have shaped and sustained the Landcare groups over the long-term.

These attributes weave together academic knowledge with the themes that emerged from this project. Three of the attributes have strong parallels with design principles that Elinor Ostrom[10] used to explain long-lasting institutions that govern common pool resources, like management of forests and groundwater resources. A further three attributes cohere with design principles that my colleagues and I identified as being important to collective weed management.[11] In the descriptions that follow, I indicate the parallels as a way of explaining how the attributes can be adopted and adapted depending on context.

Develop shared community and environment-oriented goals

All seven groups have clear goals with respect to the weeds they wanted to address and whether they are aiming for eradication or preventing spread. For example, Tathra Land Care aim to eradicate bitou bush from the Tathra area while Towamba Valley Landcare seeks to have no serrated tussock seeding, to eradicate any new invading weeds and to prevent spread of weeds to neighbouring properties. Some groups also articulate environmental or agricultural issues they wish to work on together. For example, the goal of Candelo and District Landcare is to improve native vegetation and prevent stream-bank erosion.

Past research has argued that a shared goal is important for collective management of weeds,[12] but it provides little insight into how those goals are arrived at or communicated. The groups in this book often developed their shared goals at community meetings or through preparation of grant applications and newspaper articles. While in some cases, group goals are written down and shared via newsletters or social media, the aims are mostly reaffirmed verbally. All the Landcare members I interviewed articulated the aim of their group almost verbatim, regardless of how long they had been a member.

Most of the groups also have a shared social aim. Brogers Creek Landcare is the clearest with respect to having dual environmental and social goals: they aim to protect their environment and nurture community. Other groups explicitly include education as one of their aims. For example, Upper Kangaroo Valley Landcare seeks to raise awareness about natural resource degradation and educate the community about their diverse native ecosystem. Similarly, Candelo and District Landcare aim to educate the community about the value of the creek and the need to look after it. For other groups, the social goal is implicit but evident in the way that the group is run.

Once the shared goals have been established, it is important that the group regularly reflects on whether the goals are being achieved. Regardless of whether the group has an explicit goal of nurturing community, the sharing of home-made cakes is instrumental to most groups. Cake facilitates social bonding and provides an opportunity for collective reflection on, and celebration of, achievement of group goals. The social side of Landcare is just as important as the environmental and agricultural.

Be realistic in setting the size of the area to work

All the groups have clearly defined the boundaries of where they focus their efforts, spanning from 1 km to 42 km of river, from 50 to 80 properties or ~300 ha. These diverse boundaries reflect multiple interdependent factors that determine whether the shared goals will be achievable. These factors include: the extent of the weed problem; whether one or multiple weeds are being targeted; whether the group works on public or private land; the number of people available to meet regularly; the frequency of working bees; and how much, if any, external funding is sought to pay contractors to support the work of the group. The importance of having well-defined natural and social boundaries is consistent with Elinor Ostrom's work, which highlighted the importance of clearly defining a resource and who takes responsibility for it.[13]

Some groups, like Candelo and District Landcare, Deua Rivercare and Upper Kangaroo Valley Landcare, have increased the area that they work in over time, expanding as they make progress within their defined boundaries and to address where weed seeds are coming from. This expanding of boundaries seems to be more characteristic of challenges like weeds than other forms of collective management.[14]

Beyond delimiting the areal extent of group work, all the groups subdivide the area in which they work to make the task of weeding (or planting trees or controlling pest animals) more manageable. None of the groups aim to control all weeds everywhere within their clearly defined area all the time. For some groups the subdivision is explicit and strategic. For example, Tathra Land Care has ~20 sub-areas that they visit at least once every two years to keep an eye on bitou bush. Similarly, there are eight stretches of Upper River that Upper Kangaroo Valley Landcare visit each year. Other groups, like Deua, Brogers Creek, Tapitallee Budgong or Towamba Valley, focus their effort on individual properties on an as-needs basis. The need for sub-areas to focus weeding efforts is consistent with more recent collective action research that has found it can be useful to focus on one smaller problematic area at a time rather than tackle the larger area as a whole.[15]

Beyond defining where work will take place, geographical features help to provide a sense of community identity and responsibility for weeding the clearly defined area. For example, for the Deua, Upper Kangaroo River, Brogers Creek and Towamba groups, living in an out-of-the-way valley provides a sense of attachment to others who live nearby. For the first three of these groups the river and road also provide a sense of belonging. Similarly, most of the people who live in and around Candelo either have a view of the creek or see it when they come into town. Towamba, Tathra and Deua are communities bounded by national parks and either a reserve or state forest. Being enclosed in this way contributes to the sense of belonging. Beyond providing the basis for community identity, such geographical boundaries shape the nature of the weed challenge and the likelihood that people relate to the problem being addressed.

Take every opportunity to invite people to join the group

Most groups have one person who is proactive in their efforts to recruit new members. It is nigh impossible to be in Tathra without meeting Jim and being invited to join the Landcare group. In Brogers Creek, Andrew is renowned for meeting new neighbours as soon as the removalist truck has rolled into the neighbourhood. In Towamba, Derek is known as the doorknocker, who makes sure everyone knows about serrated tussock, African lovegrass, and that they are welcome to join Landcare meetings. In Deua, Anne used her extensive networks to invite everyone in the valley to discuss their weed challenges and join the group. Being attentive to geographical features can help to identify who may be interested in joining the group.

While most groups regularly have 10 to 20 people join their working bees, such numbers only stay steady or grow over time with active recruitment efforts. Most groups work on a model where they have both active and irregular members. Their membership and email lists are much larger than the numbers of people at working bees. Upper Kangaroo Valley Landcare has a model where there are many inactive financial members, recognising that there are people who want to support the work of the group but may not be able to join working bees. This model has worked well to enable the group to pay annual costs, such as insurance, and the group only needs to request funds occasionally for specific weed projects.

Past research on collective action suggests that group size can be a challenge.[16,17,18] If a group becomes too large it can become unwieldy; if it is too small it can lose momentum. For the Landcare groups in this book, the sustainable size seemed to require having between 8 and 25 attendees at working bees.

Value all contributions

To create a culture of mutual trust and respect, it is important that everyone feels valued. Some groups are explicit that they want to be inclusive of all ages and abilities and make sure that they find roles to suit everyone's needs. Others communicate widely that all contributions are welcome, regardless of how often or regularly people can help.

As part of valuing everyone, past research has noted the importance contributions being made transparently.[19] This provides reassurance that individual efforts are worthwhile and are being reciprocated. In the case of Towamba Valley Landcare, individual efforts and commitment to the cause are demonstrated by attending meetings every 6 months where everyone reports on their progress and challenges. For others, individuals report on the number of hours that they contribute to controlling weeds outside of working bees. These hours are summed and reported at Annual General Meetings (in the case of the Upper Kangaroo Valley) or in end-of-project reports back to funding organisations (such as Brogers Creek Landcare and Deua Rivercare). Contributions are also made transparent by participation in working bees. Being present and lending a hand is noted and appreciated by group members and is reciprocated at subsequent working bees.

Valuing all contributions also means documenting, recognising and valuing what the group has achieved as a collective. This ties in with reflecting on whether the group goals are being achieved and where there is work yet to be done. Monitoring can take many forms. Some groups take photos before and after working bees or at the start and end of group projects. Others, like Deua Rivercare, map riparian weeds every second year. The data can then be compiled and presented to demonstrate the impact the group is having on the environment. It is also important to reflect on social achievements. Such reflections on what the group means to members over time, through personal stories, can also be particularly helpful when the group is at a crossroads – trying to decide whether goals need updating or whether to continue when a leader steps down. It is important to value and reflect on contributions regularly – at the end of projects, at Annual General Meetings or even when new members join the group. Celebrate successes early and often!

Share leadership and administrative responsibilities

All the groups shared responsibility for leadership. This often involved collaboratively deciding where to prioritise group efforts, communicating with members about upcoming events, recruiting new members, liaising with government and non-government organisations, applying for and managing external funding, and reinforcing the group norms and culture.

The benefit of sharing leadership is that these tasks can be allocated to people who have a particular skill or talent for these activities. Whether it be someone who 'gets on with just about everybody' being responsible for recruitment or someone who is handy with email and social media being responsible for regular communications. No single person alone is required to do all these activities. Sharing the administrative burden means that the remaining group members feel secure that things are taken care of and they feel a sense of obligation to the leadership group – they appreciate the work being done and want to continue participating as a way of showing their gratitude. This resulted in consistent, passionate and long-lasting leadership that inspires and enthuses others.

The notion of shared leadership is also important when it comes time to hand over responsibilities. None of the groups had leadership succession plans, and many members raised concerns about what would happen when the key people leading the group no longer wanted to continue. In some ways, the shared leadership model facilitates leadership succession, especially when those who have been leading a group then mentor those who are taking over, providing valuable guidance on what makes the group tick. Similarly, if multiple people offer to share the load and help the group transition to a new leadership team, it is more likely that new volunteers will step forward.

Research from other fields suggest that there are other strategies for leadership succession that might benefit Landcare groups. These include explicit inclusion of newcomers into decision-making.[20] In the case of Landcare groups, this could involve including newcomers in decisions about where and when working bees take place. Those sharing leadership can encourage others by being flexible in their approach and responding positively to new ideas. Talking openly about jobs and the need for new volunteers is also recommended. Making time at the end of working bees for reflection on the administrative contributions being made by members may help make leadership succession less daunting.

Design principles for collective action groups outlined in past research are noticeably silent on the role of leadership[21] and leadership succession. While there is agreement in past research on Landcare[22] and other local collective action initiatives[23] that leadership is important, the nature of the leadership and how transitions occur is often unclear. The results here indicate the importance of considering leadership as a shared endeavour that keeps administrative burdens on any one individual to a minimum and where the often invisible work of those sharing leadership responsibilities is also recognised.

Anticipate and manage conflict

Most of the groups reported experiencing some conflict, which is to be expected when managing natural resources[24] and in places experiencing increasing social diversity.[25] Generally, the conflict was between the group and others in the community about whether specific plants were weeds or whether the approach taken to control weeds, such as the use of herbicides, was appropriate or justifiable.

There are two main ways that the groups deal with such conflict. One way involves seeking to communicate often and openly with the broader community about why particular weeds are the focus of the group's efforts. For example, Candelo and District Landcare has invested in highly visible and informative signs near Candelo Creek that explain the geomorphology of the river system and the impacts that willows have. They have also held photo musters that invite the community to bring in photos of the creek, to engage them in a conversation about this valued place.

The other way the groups deal with conflict is by being respectful and valuing all contributions. This is particularly important for groups where there is resistance to the use of herbicides to control weeds. Both Brogers Creek Landcare and Upper Kangaroo Valley Landcare are clear that members can choose whether or not they use herbicides during working bees. Herbicides are provided in small bottles so members can cut and paint woody weeds, but even then, there is no expectation that herbicides are used. This is reassuring to members who do not want to use herbicides and feel valued despite their wish to avoid herbicides.

In both ways, groups address conflict by clearly communicating their rationales for choosing which weeds to focus on and for controlling weeds in specific ways. Such communication forms the basis of trusting relationships and identification of shared values, which allows for diverse opinions to be respectively expressed and accommodated. Groups accept that Landcare is not for everyone and that it is inevitable that tensions will arise and persist regardless of their communications and commitment to being inclusive.

Collective action design principles indicate that it is important for groups to have low cost conflict resolution mechanisms.[26] While past research has focused on formal avenues for resolving conflicts, the seven Landcare groups developed their own informal approaches for managing tensions locally and accepting that some level of conflict and tension is not only unavoidable but can also be productive.[27]

Ask for and accept help from government and non-government organisations

All the groups have received in-kind and financial support from government organisations. In-kind support has included providing advice about which weeds and what area to focus on, helping notify the broader community about priority weeds that match the aims of the group, undertaking mapping, helping with weed identification, providing access to weed control equipment, organising community barbecues, giving presentations and updates on topics of interest, and encouraging others in the community to join the Landcare group, or at least contribute to the shared goal by undertaking weed control individually.

Beyond helping Landcare groups with their own activities, public land managers provide in-kind support by prioritising and broadening their weed control activities to align with those of the group. Such in-kind support provides groups with reassurance that the work they are doing has the support of government and that their efforts dovetail with those happening at the broader scale.

Government organisations, such as local councils, catchment management authorities, local land services, the NSW Environmental Trust, as well as many state and federal government departments, have provided financial support to the Landcare groups. This has included funding to hire contractors to undertake weed or pest animal control work, paying for equipment such as spray trailers, mattocks or tree planting materials, disaster recovery activities, educational materials and activities, data collection and more. The groups have received funds, from the minimal to the substantial, depending on the scale of the work and the type of weed control being undertaken. In many cases, the funding is provided on the understanding that financial investment will be matched two- to threefold by volunteer labour.

Landcare groups are also assisted by local and regional coordinators. These coordinators support group activities by promoting Landcare in the region, helping to apply for, manage and report on government grants, and linking Landcare groups together across a region so that they can learn from one another. The support provided by Landcare coordinators and government agencies is consistent with collective action design principles that identify the importance of groups having broader institutional support.[28]

Overall, the stories in this book speak to the value of Landcare as a movement that has lasted 40 years thanks to its adaptability and stability. The flexibility of the Landcare model is key. It enables local groups to tailor the organisational structure to their own needs – adapting it to the people, the place, the plants. While the seven attributes identified above address similarities across the groups, these attributes were rarely applied in the same way by more than one group. The groups have autonomy over their goals and activities, the leadership structure, whether or not they seek external funding, and the locations and frequency with which they meet.

The regional, state and national Landcare organisations also provide long-term stability thanks to bipartisan support over decades. The persistence of the organisations has enabled local and regional Landcare coordinators to support groups establish, define goals and manage funding. For some groups, this support makes all the difference. Not having to undertake all the administrative work frees them up to spend more time undertaking activities that they are passionate about. For other groups, this stability helps with leadership succession – the longevity of Landcare provides support when leadership changes hands.

While Landcare provides long-term continuity that supports on-ground works, it is also important to note that the amount of funding it receives has varied significantly over time. Some have highlighted how this hampers the ability for the model to reach its full potential.[29] Consistent ongoing funding is needed to ensure that Landcare groups are supported at district, regional, state and national levels. This will enable groups to continue to provide community support and resilience through the unprecedented changes we are facing today.

An invitation to join Landcare

At the start of this book, I highlighted that weeds are everywhere and that my Aboriginal interviewee wanted to know: who is going to come and help care for Country?

Progress has been made. Seven Landcare groups were started over the last three decades by people wanting to make a difference to their local environment. At a time when discourse on climate change and other world events can make us despair, the difference these seven Landcare groups have made is remarkable. By working together – often no more than a dozen people, regularly and over time – these groups have cared for the highly modified landscapes that surround us.

Collectively, these groups have removed weeds one at a time to enable other plants to take their place, changing the way fires and floods move across the landscape and how animals and people relate to Country.

What the stories in this book show is that Landcare is effective when it nurtures social connections. Joining Landcare can provide an opportunity to learn together about the environment and establish new relationships with neighbours.

There is also something about weeding together that encourages people to open up about their challenges, share emotions and provide one another with practical support. The cake, tea and coffee afterwards cement friendships and provide an opportunity to celebrate what has been achieved.

To make a difference, I encourage you to find out where your local Landcare, Rivercare, Bushcare, Dunecare or other 'care' group meets and join them for a working bee. The National Landcare Directory[30] is a good place to start, although some of the groups in this book aren't listed, so if in doubt, contact your Local Landcare coordinator.[31]

If there are no groups in your area, be brave and start a conversation with your neighbours about weeds. Long-term residents have a depth of knowledge and social networks. Local Landcare coordinators can also help; they might be able to put you in touch with other landholders or support you to gauge interest in starting a group.

If you're already a member of a Landcare or other 'care' group, I hope you feel invigorated to invite someone new to come along to a Landcare event and experience the difference it can make. And don't forget to bring the next generation with you – your children, nieces, nephews, grandchildren. My daughter and her best friend love it when I take them with me – a chance to get out, make new friends, pull out weeds and plant trees.

Now it's time to address weeds together. Weeding can make a difference to local environments, their susceptibility to fires and floods, the continuity of habitats for threatened species and the ability to grow productive crops. Weeding, when done collectively, can change the resilience of landscapes and strengthen the communities who live within them. The reality and likely future effects of climate change are daunting, yet the stories in this book show that there is something real and tangible that we can all do. So let's go weeding (and eat cake)!

Recipe: Cherry Ripe Slice

It would be remiss of me to not include a slice recipe in this book. Slices are a popular choice at working bees, like this classic from Tuncurry Dunecare.

Ingredients

Base

2 cups self-raising flour
2 cups brown sugar
2 cups coconut
1 cup melted butter or margarine

Cherry filling

1 tin sweetened condensed milk
200 g finely chopped glacé cherries
125 g melted copha
1½ cups desiccated coconut
A few drops red food colouring

Chocolate topping

60 g copha
375 g dark or milk chocolate

Method

1. Preheat oven to 160°C.
2. Line two 23 × 23 cm tins with baking paper.
3. Combine all ingredients for the base into a large bowl and mix well.
4. Press half of the mixture into each tin, until the base is flat and firm.
5. Bake bases in oven for 15 minutes, then remove and allow to cool in the tin.
6. Mix the cherry filling ingredients together in a large bowl and then split it in half and spread over the cooled bases with a knife.

7. Melt the copha and add the chocolate pieces until they are all melted.
8. Spread the melted chocolate over the slice.
9. Allow to set at room temperature or in the fridge for a crisper top.

Jean Atchison was known for her excellent baking that was highly appreciated by Tuncurry Dunecare, of which she and her husband, Frank Atchison, were founding members. According to the NSW Legislative Assembly Parliamentary Debates (Hansard) 14 October 2021, the morning tea feasts Jean provided were essential to ensuring that this group was successful. This recipe was the favourite of her granddaughter, Jenny Atchison.

ACKNOWLEDGMENTS

This book has been a highly collaborative endeavour.

I want to begin by thanking all the members of the Landcare groups (see List of Participants for those happy to be named) who volunteered to be part of the research. I was always made to feel welcome at working bees. I am so grateful for the all the local knowledge shared and the time spent explaining what it is like to be a member of a Landcare group. Special thanks to the people who shared their baked cakes, slices and biscuits; your contributions have nourished me throughout this project.

A special thank-you to the people who convene the Landcare groups, for facilitating access to the groups, letting me know when events were happening and for providing feedback on the chapters. Your dedication inspires me.

I am grateful for the Local and Regional Landcare coordinators, who helped me to make contact with the Landcare groups at the start of this research journey and who have helped me along the way. They have invited me to come along to district and regional Landcare meetings, have read the chapters of this book and have joined me in various outreach activities over the years.

The South East Landcare network has embraced me as one of their own; I spent three years on the South East Landcare committee. I learned much about the organisation of Landcare from taking on this role and enjoyed working with this passionate and highly motivated group of volunteers.

This research would not have been possible without the support of my fabulous colleagues at the Australian Centre for Culture, Environment, Society and Space (ACCESS) at the University of Wollongong. Before I even submitted the application, Nick Gill, Pauline McGuirk and Gordon Waitt, helped my ideas take shape and supported me in navigating the internal workings of applying for grants at the university. It is without doubt that my success in being awarded an

Australian Research Council Discovery Early Career Researcher Award (DE200100234), which funded this project, is thanks to their insightful feedback and support.

Joining the then School of Geography and Sustainable Communities and working with the sensational team of human geographers at the University of Wollongong has been an absolute joy and highlight of my academic career. I am in awe of all my colleagues, their research and commitment to teaching. I name them all here because each of them has shaped this work and been part of this project in various ways: Crystal Arnold, Jenny Atchison, Thomas Birtchnell, Chris Brennan-Horley, Chantel Carr, Vanessa Cavanagh, Nicole Cook, Olivia Dun, Leah Gibbs, Laura Hammersley, Amelia Hine, Natascha Klocker, Elyse Stanes, Andrew Warren and Isabelle Wolf. Special thanks are due to Chris Gibson for encouraging me to write this book. His enthusiasm for writing books is infectious and gave me the confidence to give it a go. I appreciate all the advice he gave me on the publishing process and asking me about how it was going.

I have also been fortunate to have amazing research assistants. Gina Hawkes helped with so many things, among them finding people to interview and undertaking interviews. She provided feedback on early drafts of the Landcare chapters, facilitated a workshop in Cairns and helped me with countless other research projects and consultancies. Words can't express all my gratitude for Gina's work and support. I have also been privileged to have Kaitlyn Height help. Her expertise on social norms and collective weed management have been invaluable in helping track down relevant literature and writing papers together. We have been working together since 2013. I hope this partnership continues for many years more.

The Weedos @ Wollongong group that Jenny, Nick and I founded a few years ago has also been a delight. Thanks to Crystal, Diana Dawson, Gina, Janine Bailey, Jen Smart, Kaitlyn, Laura Butler, Marc Llewellyn, Rachel Jefferson, Rebecca Campbell and Ty Challand. What a treat to have such a great group of weedy social scientists to talk shop with. Special thanks to my PhD students, Diana, Janine and Jen – I have learnt so much more about the social side of weeds from working with you on your projects. A special thanks to Janine for your artistic advice on my botanical drawings.

Thank-you to Anthony McKnight and Crystal for supporting me in my journey of learning about Country and my relationship to it.

Thank-you to Teaniel Mifsud for producing the map of the case study region.

Thank-you to Mike Williams, Elyse Stanes and Pamela Cook for your thoughtful feedback on my manuscript. This book is better for your insights.

Thank-you to Elinor Pickard for inspiring me to include botanical drawings in my book. I have greatly enjoyed attending your botanical art classes at the Sydney Community College and am grateful for the feedback you have provided on the artwork included in this book. I have learned so much from you and look forward to continuing to develop my newfound love of drawing.

Thank-you to Emma Phillips for using your photographic expertise and flair to digitise my botanical drawings. It was fun to work together on something creative. Let's do it again sometime.

Thank-you to Caroline Lenette for always being willing to talk shop about writing books. I am in awe at the ease with which you write books and appreciate your generosity in sharing your experiences with me.

Thanks to my Dad for helping me, as he always has, to work out how to write the introduction better (fewer tears this time than during uni!) and for helping me figure out what was out of whack with the concluding chapter.

Thanks to my Mum for joining me on a field trip to Wattamolla, reading an early draft about Brogers Creek Landcare and listening to me read my chapter on Upper Kangaroo Valley Landcare when we visited Upper River together. Your dedication to writing your own book gave me confidence to give this a go and follow in your footsteps. We miss you so much.

Thanks to my husband, Warren, for supporting me throughout this research project, for reading draft chapters, listening to presentations, and hearing me go on and on about the Landcare groups. I am forever grateful to have met you and for your love and support.

Thanks to my daughter, Zoe, for joining me on working bees, planting trees, showing me how the whirligig works and for bringing joy to my life every day. I hope that one day you'll start your own Landcare group.

Finally, I would like to thank my editor and development editor, George Knott and Mark Hamilton at CSIRO Publishing for working with me on this manuscript over the last two years. I have appreciated your guidance and responses to my many questions as a first-time author. Thanks also to Janet Walker for her thoughtful copy editing.

LIST OF PARTICIPANTS

A.B. Waddell
Adam Acworth
Alison Walsh
Allan Rees
Amanda Gale
Andrew Fitzsimons
Anne Rault
Barbara Acworth
Barry Woolacott
Belinda Booth
Belinda Watson
Bridget Sach
Bruce Hamilton
Cam Johnstone
Catriona Moore
Chris Hamilton
D. Lear
Dan Crowley
Dave Darlington
David Brinson
David Fletcher
David Gallan
David Howard
Derek Lewis
Diana Richards
Don Firth
Emma Patyus
Faye Lewis
Fiona Firth
Garry Daly
Grant Forsdick

Greg Madden
Greg Thompson
Heath Mason
Howard Charles
Ian Gamble
Jacqui Williams
Jamal Airey
Jamie Dixon-Keay
Jan Phillips
Janet Bundey
Jean Bentley
Jennifer Bateman
Jessica Dwyer
Jim Kelly
John Kubale
John Paine
Julie Hilditch
Julie Ward
Karen Joynes
Ken McLeod
Kim Touzel
Lauren Van Dyke
Leon Wallace
Leonard White
Liam Ryan
Liz Cuninghame
Marien Stark
Mark Bugden
Mark Chaplin
Mark Darby
Mary Grealy
Megan Lorimer
Mel Wilkerson
Michael Andrews
Neil Brown

Nicholas Carlile
Paul Martin
Peter Smith
Peter Turner
Philippa Street
Richard Barcham
Robyn Stacey
Sarah Russell
Scott Keyworth
Shani Keane
Sonya Reyes
Stephen Wilson
Steve Taylor
Stuart Cameron
Tess Heighes
Trish Kenny
Wendy Calkhoven
Wendy Jones
Xavier Smith
Yuin Kelly
Zoe

ENDNOTES

Chapter 1

[1] Invasive Plants and Animals Committee (2016) 'Australian Weeds Strategy 2017 to 2027'. Australian Government Department of Agriculture and Water Resources, Canberra, ACT.

[2] Department of Environment and Heritage (2025) *Orange hawkweed*. NSW Department of Environment and Heritage website, accessed 30 May 2025. <https://www.environment.nsw.gov.au/topics/animals-and-plants/pest-animals-and-weeds/weeds/new-and-emerging-weeds/orange-hawkweed#weed-eradication-detector-dogs>

[3] Miller G, Friedel M, Adam P, Chewings V (2010) Ecological impacts of buffel grass (Cenchrus ciliaris L.) invasion in central Australia – does field evidence support a fire-invasion feedback? *The Rangeland Journal* **32**, 353–365.

[4] Kulitja R (3 March 2024) 'Rene Kulitja on how buffel is affecting country' [video], *Central Land Council*, YouTube, accessed 2 December 2024. <https://www.youtube.com/watch?v=sKQTwQIKjXw>. See also the 2021 Desert IPA Rangers Statement on buffel grass (The Umuwa Statement) and the 2022 Statement about buffel grass by Uluru-Kata Tjuta, Katiti-Petermann and Watarrka: https://assets.nationbuilder.com/aridlands/pages/4050/attachments/original/1698800679/Buffel_grass_weed_declaration_in_the_Northern_Territory_guide.pdf?1698800679

[5] Williams KJ, Hunter B, Schmidt RK, Woodward E, Cresswell ID (2021) Land: Introduced and invasive species. In *Australia State of the Environment 2021*. Australian Government Department of Agriculture, Water and the Environment, Canberra, ACT. <https://soe.dcceew.gov.au/land/pressures/introduced-and-invasive-species>

[6] Stenekes N, Ticehurst J, Arthur T (2024) 'Pest Animal & Weed Management Survey 2016/2019/2022'. Australian Bureau of Agricultural and Resource Economics and Sciences, Canberra, ACT.

[7] Schrader S, Graham S, Campbell R, Height K, Hawkes G (2024) Grower attitudes toward area-wide management of cropping weeds in Australia. *Land Use Policy* **137**, 107001.

[8] Graham S (2012) Social relations and natural resource management: The significance of trust and power to solving a collective weed management problem. PhD thesis. Charles Sturt University, Wagga Wagga, NSW, p. 218.

[9] Graham S (2012) Social relations and natural resource management: The significance of trust and power to solving a collective weed management problem. PhD thesis. Charles Sturt University, Wagga Wagga, NSW, p. 218.

[10] Schrader S, Graham S, Campbell R, Height K, Hawkes G (2024) Grower attitudes toward area-wide management of cropping weeds in Australia. *Land Use Policy* **137**, 107001.

[11] Schrader S, Graham S, Campbell R, Height K, Hawkes G (2024) Grower attitudes toward area-wide management of cropping weeds in Australia. *Land Use Policy* **137**, 107001.

[12] Graham S (2013) Three cooperative pathways to solving a collective weed management problem. *Australasian Journal of Environmental Management* **20**, 116–129.

[13] Invasive Species Council (2014) 'Weeds and Climate Change', Invasive Species Council, Carlton, Victoria. <https://invasives.org.au/wp-content/uploads/2014/02/fs_weedsandclimatechange.pdf>

[14] Ramesh K, Matloob A, Aslam F, Florentine SK, Chauhan BS (2017) Weeds in a changing climate: vulnerabilities, consequences, and implications for future weed management. *Frontiers in Plant Science* **8**, 95.

[15] Berry ZC, Wevill K, Curran TJ (2011) The invasive weed Lantana camara increases fire risk in dry rainforest by altering fuel beds. *Weed Research* **51**, 525–533.

[16] Arnold C, Atchison J, McKnight A (2023) 'Often in between': Thinking through research methods and Indigenous sovereignty with Yuin Country. *Environment and Planning F* **2**, 163–179.

[17] Bach TM, Kull CA, Rangan H (2019) From killing lists to healthy country: Aboriginal approaches to weed control in the Kimberley, Western Australia. *Journal of Environmental Management* **229**, 182–192.

[18] Similar views were expressed by Girringun Elders, Rangers, nursery staff in research undertaken by Maclean and colleagues. In that research, concern was expressed about large infestations of weeds that affected sacred and special places, songlines and totems. See Maclean K, Robinson C, Bock E, Rist P (2022) Reconciling risk and responsibility on Indigenous country: Bridging the boundaries to guide knowledge sharing for cross-cultural biosecurity risk management in northern Australia. *Journal of Cultural Geography* **39**, 32–54.

[19] Height K (2018) The role of social norms in collective action for weed management across rural private property. PhD thesis, University of Melbourne, Melbourne.

[20] Tonkin v Cooma-Monaro Shire Council [2006] NSWCA 50 (7 April 2006).

[21] Wild Matters (2020) *National Established Weed Priorities – Towards a National Framework*. Wild Matters, Victoria. <https://wildmatters.com.au/wp-content/uploads/2021/08/Attachment-A-National-Established-Weed-Priorities-Towards-a-National-Framework.pdf>

[22] Weeds Australia (n.d.) *Learn about the weeds you want to manage*. Weeds of National Significance website, accessed 30 May 2025. <https://weeds.org.au/weeds-profiles/>

[23] Commonwealth Department of Agriculture, Fisheries and Forestry (2025) *National Established Weeds Priorities Framework (NEWP)*, Department of Agriculture, Fisheries and Forestry website, accessed 5 June 2025. <https://www.agriculture.gov.au/biosecurity-trade/pests-diseases-weeds/pest-animals-and-weeds/established-weeds/newp>

[24] Schrader S, Graham S, Campbell R, Height K, Hawkes G (2024) Grower attitudes toward area-wide management of cropping weeds in Australia. *Land Use Policy* **137**, 107001.

[25] Campbell R, Height K, Hawkes G, Graham S, Schrader S, Blessington L, McKinnon S (2023) Meanings, materials and competences of area-wide weed management in cropping systems. *Agricultural Systems* **212**, 103776.

[26] Marshall GR, Coleman MJ, Sindel BM, Reeve IJ, Berney PJ (2016) Collective action in invasive species control, and prospects for community-based governance: The case of serrated tussock (Nassella trichotoma) in New South Wales, Australia. *Land Use Policy* **56**, 100–111.

[27] Landcare Australia (2025) *Origins of Landcare*, Landcare Australia website, accessed 30 May 2025. <https://landcareaustralia.org.au/about-us/origins-of-landcare/>

[28] Campbell A (1994) *Landcare: communities shaping the land and the future*. Allen and Unwin, Sydney, NSW.

[29] Youl R, Marriott S, Nabben T (2006) *Landcare in Australia founded on local action*. Secretariat for International Landcare and Rob Youl Consulting Pty Ltd, Victoria.

[30] Department of Agriculture, Fisheries and Forestry (2022) *Landcare*. Commonwealth Department of Agriculture, Fisheries and Forestry website, accessed 2 December 2024. <https://www.agriculture.gov.au/

agriculture-land/farm-food-drought/natural-resources/landcare#:~:text=There%20are%20around%206%2C000%20groups,Australia%20caring%20for%20the%20land>

[31] Tennent R, Lockie S (2013) Vale Landcare: the rise and decline of community-based natural resource management in rural Australia. *Journal of Environmental Planning and Management* **56**, 572–587.

[32] Munday B (2017) *Those wild rabbits: How they shaped Australia*. Wakefield Press, South Australia.

[33] Robins L, Kanowski P (2011) 'Crying for our Country': eight ways in which 'Caring for our Country' has undermined Australia's regional model for natural resource management. *Australasian Journal of Environmental Management* **18**, 88–108.

[34] Campbell A (2009) It's time to renew Landcare. *Agricultural Science* **2**, 30–33.

[35] Biddle N, Boyer C, Gray M, Jahromi M (2022) 'Volunteering in Australia: The Volunteer Perspective.' Volunteering Australia, accessed 3 February 2025. <https://volunteeringhub.org.au/wp-content/uploads/2024/06/Volunteering-in-Australia-2022-The-Volunteer-Perspective.pdf>

Chapter 2

[1] Burarrwanga L, Ganambarr R, Ganambarr-Stubbs M, Ganambarr B, Maymuru D, Wright S, Suchet-Pearson S, Lloyd K (2013) *Welcome to my Country*. Allen and Unwin, Melbourne, Victoria.

[2] Country B, Wright S, Suchet-Pearson S, Lloyd K, Burarrwanga L, Ganambarr R, Maymuru D, Sweeney J (2016) Co-becoming Bawaka: Towards a relational understanding of place/space. *Progress in Human Geography* **40**, 455–475.

[3] Nursey-Bray M, Palmer R, Stuart A, Arbon V, Rigney LI (2020) Scale, colonisation and adapting to climate change: Insights from the Arabana people, South Australia. *Geoforum* **114**, 138–150.

[4] Burarrwanga L, Ganambarr R, Ganambarr-Stubbs M, Ganambarr B, Maymuru D, Wright S, Suchet-Pearson S, Lloyd K (2013) *Welcome to my Country*. Allen and Unwin, Melbourne, Victoria.

[5] Australian Institute of Aboriginal and Torres Strait Islander Studies (2022) *Welcome to Country*, AIATSIS website, accessed 19 May 2025. <https://aiatsis.gov.au/explore/welcome-country>

[6] Reyes-García V, Arnold C, Graham S (2024) Indigenous Peoples provide alternative approaches to managing biological invasions. *Trends in Ecology & Evolution* **39**, 790–792.

[7] Burarrwanga L, Ganambarr R, Ganambarr-Stubbs M, Ganambarr B, Maymuru D, Wright S, Suchet-Pearson S, Lloyd K (2013) *Welcome to my Country*. Allen and Unwin, Melbourne, Victoria.

[8] Arnold C (unpublished) How can we re-envision care for weeds? Indigenous weed management on the Shoalhaven River. ACCESS Seminar, University of Wollongong, Wollongong, 19 March 2025.

[9] Cavanagh V, Standley P (2020) Walking in the landscapes of our ancestors-indigenous perspectives critical in the teaching of geography. *Interaction* **48**, 14–16.

[10] McKnight A (2016) Meeting Country and self to initiate an embodiment of knowledge: Embedding a process for Aboriginal perspectives. *The Australian Journal of Indigenous Education* **45**, 11–22, p. 12.

[11] McKnight A (2016) Meeting Country and self to initiate an embodiment of knowledge: Embedding a process for Aboriginal perspectives. *The Australian Journal of Indigenous Education* **45**, 11–22.

[12] Arnold C, Atchison J, McKnight A (2023). 'Often in between': Thinking through research methods and Indigenous sovereignty with Yuin Country. *Environment and Planning F* **2**, 163–179.

[13] Burarrwanga L, Ganambarr R, Ganambarr-Stubbs M, Ganambarr B, Maymuru D, Wright S, Suchet-Pearson S, Lloyd K (2013) *Welcome to my Country*. Allen and Unwin, Melbourne, Victoria.

[14] Country B, Wright S, Suchet-Pearson S, Lloyd K, Burarrwanga L, Ganambarr R, Maymuru D, Sweeney J (2016) Co-becoming Bawaka: Towards a relational understanding of place/space. *Progress in Human Geography* **40**, 455–475.

[15] McKnight A (2016) Meeting Country and self to initiate an embodiment of knowledge: Embedding a process for Aboriginal perspectives. *The Australian Journal of Indigenous Education* **45**, 11–22, p. 13.

[16] McKnight A (2016) Meeting Country and self to initiate an embodiment of knowledge: Embedding a process for Aboriginal perspectives. *The Australian Journal of Indigenous Education* **45**, 11–22.

[17] Arnold C, Atchison J, McKnight A (2023). 'Often in between': Thinking through research methods and Indigenous sovereignty with Yuin Country. *Environment and Planning F* **2**, 163–179.

[18] McKnight A (2016) Meeting Country and self to initiate an embodiment of knowledge: Embedding a process for Aboriginal perspectives. *The Australian Journal of Indigenous Education* **45**, 11–22.

[19] A soft charcoal rubbing of the casuarina bark puzzle inspired by Janine Bailey's Conversations with trees project. Bailey J (2023) *Conversations with trees*, Janine Bailey website, accessed 19 May 2025. <https://www.janinebailey.com/conversations-with-trees/>

[20] Arnold C, Atchison J, McKnight A (2023). 'Often in between': Thinking through research methods and Indigenous sovereignty with Yuin Country. *Environment and Planning F* **2**, 163–179. p. 175.

[21] Graham S (7 May 2024) 'This group rid one Australian river of its privet problem – and strengthened community along the way'. *The Conversation*, accessed 5 June 2025. <https://theconversation.com/this-group-rid-one-australian-river-of-its-privet-problem-and-strengthened-community-along-the-way-226230>

Chapter 3

[1] Organ M (1990) *Illawarra and South Coast Aborigines 1770 to 1850*. Aboriginal Education Unit, The University of Wollongong, Wollongong, NSW.

[2] Brogers Creek Landcare Group (2011) *Views and Visions: Stories of Wattamolla and Woodhill*. Brogers Creek Landcare, Wattamolla, New South Wales, p. 11.

[3] Organ M (1990) *Illawarra and South Coast Aborigines 1770 to 1850*. Aboriginal Education Unit, The University of Wollongong, Wollongong, NSW.

[4] Brogers Creek Landcare Group (2011) *Views and Visions: Stories of Wattamolla and Woodhill*. Brogers Creek Landcare, Wattamolla, New South Wales, p. 116.

[5] Brogers Creek Landcare Group (2011) *Views and Visions: Stories of Wattamolla and Woodhill*. Brogers Creek Landcare, Wattamolla, New South Wales, p. 241.

[6] Australian Bureau of Statistics (2021) *Wattamolla 2021 Census All persons QuickStats*, Australian Bureau of Statistics website, accessed 5 June 2025. <https://www.abs.gov.au/census/find-census-data/quickstats/2021/SAL14207>

[7] Daly G (2012) *Rehabilitating Shoalhaven Landscapes*. Shoalhaven Landcare Inc., Tapitallee, New South Wales, p. 23.

[8] Jane P (1995) Ten of the worst. *Sydney Morning Herald*, 26 March, p. 40.

[9] Sewell L (1998) Privet: A 'health hazard'. *Illawarra Mercury*, 11 March, p. 21.

[10] Brogers Creek Landcare Group (2011) *Views and Visions: Stories of Wattamolla and Woodhill*. Brogers Creek Landcare, Wattamolla, New South Wales, p. 11.

[11] FitzSimons A (2010) Team Wattamollans issue a challenge for the 125th Kangaroo Valley Show. *Kangaroo Valley Voice*, February, p. 3.

[12] Thompson G (2005) Madeira vine infestations to be attacked. *Kangaroo Valley Voice*, July, p. 19.

[13] Thompson G (2005) Madeira vine infestations to be attacked. *Kangaroo Valley Voice*, July, p. 19.

[14] Weeds Australia (2020) *Anredera cordifolia (Ten.) Steenis: Madeira vine*. Centre for Invasive Species Solutions, Canberra. <https://profiles.ala.org.au/opus/weeds-australia/profile/Anredera%20cordifolia>

[15] School of Literature, Language and Linguistics (2017) *Meanings and origins of Australian words and idioms*. Australian National University, Canberra. <https://slll.cass.anu.edu.au/centres/andc/meanings-origins/y>

[16] Height K (2018) The role of social norms in collective action for weed management across rural private property. PhD thesis, University of Melbourne, Melbourne.

[17] Thompson G (2019) Brogers Creek Landcare. *Kangaroo Valley Voice*, August, p. 26.

[18] Ostrom E (1990) *Governing the Commons: The Evolution of Institutions for Collective Action*. Cambridge University Press, New York, USA, p. 90.

[19] Cox M, Arnold G, Tomás SV (2010) A review of design principles for community-based natural resource management. *Ecology and Society*, **15**.

[20] Walsh A, Kingwell R (2021) Economic implications of the loss of glyphosate and paraquat on Australian mixed enterprise farms. *Agricultural Systems* **193**, 103207.

[21] Ostrom E (1990) *Governing the Commons: The Evolution of Institutions for Collective Action*. Cambridge University Press, New York, USA, p. 90.

[22] Cox M, Arnold G, Tomás SV (2010) A review of design principles for community-based natural resource management. *Ecology and Society*, **15**.

Chapter 4

[1] 'Foot-bridge', *The Nowra Leader*, 19 October 1934, p. 8.

[2] For example, 'A foot suspension bridge over the Upper Kangaroo River (North Arm)', *Sydney Mail*, 30 December 1925, p. 41

[3] 'In the beautiful Kangaroo Valley', *Sydney Mail*, 24 March 1926, p. 5.

[4] 'Rolling hills in a river valley', *Australian Women's Weekly*, 4 November 1981, p. 78

[5] Visit Kangaroo Valley (2025) *Upper Kangaroo River Road*. Kangaroo Valley Chamber of Tourism and Commerce, Kangaroo Valley, NSW. <https://visitkangaroovalley.com.au/listing/upper-kangaroo-river-road/>

[6] Shoalhaven City Council differentiates between Bushcare groups that work on public property and Landcare groups that work on private property. This distinction is common in other council areas, but this distinction is not uniformly made. For example, Tathra and Candelo and District Landcare groups work on public land in the Bega Valley Shire Council area.

[7] 'Tourism changing the valley's face', *Canberra Times*, 28 August 1988, p. 16.

[8] Coombs R (2014) *Shoalhaven District Pictorial History*. Kingsclear Books, Alexandria, NSW.

[9] 'Monthly events', *Kangaroo Valley Voice*, November 2005, p. 31.

[10] Ostrom E (1990) *Governing the Commons: The Evolution of Institutions for Collective Action*. Cambridge University Press, New York, USA, p. 90.

[11] Cox M, Arnold G, Tomás SV (2010) A review of design principles for community-based natural resource management. *Ecology and Society*, **15**.

[12] Graham S, Metcalf AL, Gill N, Niemiec R, Moreno C, Bach T, Ikutegbe V, Hallstrom L, Ma Z, Lubeck A (2019) Opportunities for better use of collective action theory in research and governance for invasive species management. *Conservation Biology* **33**, 275–287.

[13] 'How green is their valley', *Sydney Morning Herald*, 14 January 2008 (online). <https://www.smh.com.au/environment/how-green-is-their-valley-20080114-gdrwo8.html>

[14] NSW Office of Environment and Heritage (n.d.) *Spotted-tailed quoll – profile*. NSW Office of Environment and Heritage webpage, accessed 30 May 2025. <https://threatenedspecies.bionet.nsw.gov.au/profile?id=10207>

[15] McKinnon S, Green A, Daly M, Cooper P, Eriksen C, Boehme T (2022) 'Bushfire research Brief: Building community resilience to bushfires – a case study of Kangaroo Valley'. University of Wollongong, Wollongong, NSW. <https://documents.uow.edu.au/content/groups/public/@web/@raid/documents/doc/uow270934.pdf>

[16] Fernandez T (2022) 'Kangaroo Valley picks up the pieces after roads closed following yet another natural disaster', *ABC News*, 8 July (online).

<https://www.abc.net.au/news/2022-07-09/kangaroo-valley-consecutive-natural-disasters-roads-closed/101220266>

[17] Crivellaro G (2022) Upper Kangaroo River community's resilience shines as major landslips cut off access, *South Coast Register*, 14 April (online).

[18] Pizarro J (2022) *The Spanish Home Kitchen*. Hardie Grant Books, London UK.

Chapter 5

[1] Australian Tapestry Workshop (n.d.) *Great Hall Tapestry*, Australian Tapestry Workshop website, accessed 4 December 2024. <https://austapestry.com.au/content/great-hall-tapestry>

[2] BirdLife Shoalhaven (n.d.) *The Shoalhaven*, BirdLife Shoalhaven website, accessed 4 December 2024. <https://www.birdlifeshoalhaven.org/shoalhaven.html>

[3] NSW National Parks & Wildlife Service (2016) 'Bugong National Park and Tapitallee Nature Reserve'. Office of Environment and Heritage, Sydney, NSW.

[4] Daly G (2024) 'Threatened flora and fauna of Tapitallee. A biodiverse suburb in the Shoalhaven Local Government Area'. Shoalhaven Landcare, Nowra, NSW.

[5] Daly G (2022) Changes to the status of birds in the Tapitallee region on the south coast of New South Wales from 1928 to 2020. *Australian Zoologist* **42**, 30–54.

[6] Daly G (2024) 'Threatened flora and fauna of Tapitallee. A biodiverse suburb in the Shoalhaven Local Government Area'. Shoalhaven Landcare, Nowra, NSW.

[7] Shoalhaven Landcare (30 May 2023) 'Bengalee Landcare Lantana Clearing Project' [video], Shoalhaven Landcare, YouTube, accessed 4 December 2024. <https://www.youtube.com/watch?v=LarPWXk23WE>

[8] Centre for Invasive Species Solutions (2021) *Lantana, Common Lantana, Kamara Lantana, Large-leaf Lantana, Pink Flowered Lantana, Red Flowered Lantana, Red-Flowered Sage, White Sage, Wild Sage* Lantana camara L. Weeds Australia website, accessed 4 December 2024. <https://weeds.org.au/profiles/lantana-common-kamara/>

[9] Ensbey R (2008) Lantana. *Primefacts: profitable and sustainable primary industries*, 673, 1–8. <https://www.dpi.nsw.gov.au/__data/assets/pdf_file/0008/256472/Lantana.pdf>

[10] Centre for Invasive Species Solutions (2021) *Lantana, Common Lantana, Kamara Lantana, Large-leaf Lantana, Pink Flowered Lantana, Red Flowered Lantana, Red-Flowered Sage, White Sage, Wild Sage* Lantana camara L. Weeds Australia website, accessed 4 December 2024. <https://weeds.org.au/profiles/lantana-common-kamara/>

[11] 'Noxious Weeds', *Shoalhaven Telegraph*, 8 August 1888, p. 2.

[12] 'Nowra Municipal Council', *Shoalhaven Telegraph*, 26 September 1888, p. 2.

[13] 'Noxious weeds', *Nowra Leader*, 26 February 1932, p. 6.

[14] Centre for Invasive Species Solutions (2021) *Lantana, Common Lantana, Kamara Lantana, Large-leaf Lantana, Pink Flowered Lantana, Red Flowered Lantana, Red-Flowered Sage, White Sage, Wild Sage* Lantana camara L. Weeds Australia website, accessed 4 December 2024. <https://weeds.org.au/profiles/lantana-common-kamara/>

[15] Shoalhaven Landcare (30 May 2023) 'Bengalee Landcare Lantana Clearing Project' [video], Shoalhaven Landcare, YouTube, accessed 4 December 2024. <https://www.youtube.com/watch?v=LarPWXk23WE>

[16] Pickard N (2019) Bengalee Landcare invites community to field day. *South Coast Register*, 9 April (online).

[17] Shoalhaven Landcare (30 May 2023) 'Bengalee Landcare Lantana Clearing Project' [video], Shoalhaven Landcare, YouTube, accessed 4 December 2024. <https://www.youtube.com/watch?v=LarPWXk23WE>

[18] NSW Rural Fire Service (2019) 'Planning for Bush Fire Protection 2019: A guide for councils, planners, fire authorities and developers'. NSW Rural Fire Service, Granville, NSW, p. 87.

[19] Bundanon (2025) *Art Museum*, Bundanon, Illaroo, NSW. <https://www.bundanon.com.au/art-museum/>

[20] Note that Landcare Australia has few direct ties to local Landcare groups. The National Landcare Network is the peak body that represents community Landcare groups at the national level. Landcare NSW (2017) *New national body one step closer*, Landcare NSW website, accessed 30 May 2025. <https://landcarensw.org.au/new-national-body-one-step-closer/>

[21] Bundanon Trust (2024) 'Bundanon: Annual Report 2023–2024'. Bundanon Trust, North Nowra, NSW.

[22] *Bengalee Landcare Newsletter*, February 2023, p. 2.

[23] Dunn R (2019) Budding partnerships for bird conservation. *Birdlife Shoalhaven Magazine* 22, p. 10.

Interlude: Engaging amenity migrants

[1] Abrams JB, Gosnell H, Gill NJ, Klepeis PJ (2012) Re-creating the rural, reconstructing nature: An international literature review of the environmental implications of amenity migration. *Conservation and Society* **10**, 270–284.

[2] Wright S (2022) How many city slickers used COVID for sea and tree change. *Sydney Morning Herald*, 27 July (online). <https://www.smh.com.au/politics/federal/city-slickers-use-covid-for-sea-and-tree-change-20220726-p5b4r7.html>

[3] Gill N, Lewis A, Chisholm L (15 June 2023) 'Are Tree Changers Bad at Managing their Properties? A new study wades into the weeds to find the answer'. *The Conversation*, accessed 24 February 2025. <https://theconversation.com/are-tree-changers-bad-at-managing-their-rural-properties-a-new-study-wades-into-the-weeds-to-find-the-answer-206410>

[4] Graham S (2012) Social relations and natural resource management: the significance of trust and power to solving a collective weed management problem. PhD Thesis, Charles Sturt University, Wagga Wagga, NSW.

[5] Klepeis P, Gill N, Chisholm L (2009) Emerging amenity landscapes: invasive weeds and land subdivision in rural Australia. *Land Use Policy*, **26**, 380–392.

[6] Gill N, Lewis A, Chisholm L, Adan N (2024) What is the problem with absentee landowners? Invasive plant management by residential and absentee amenity rural landowners. *Journal of Environmental Planning and Management* **67**, 3097–3117.

[7] Height K (2018) The role of social norms in collective action for weed management across rural private property. PhD Thesis, University of Melbourne, Melbourne, Victoria.

[8] Graham S (2012) Social relations and natural resource management: the significance of trust and power to solving a collective weed management problem. PhD Thesis, Charles Sturt University, Wagga Wagga, NSW.

[9] Adie BA (2020) Place attachment and post-disaster decision-making in a second home context: A conceptual framework. *Current Issues in Tourism* **23**, 1205–1215.

[10] Landcare NSW (n.d.) *Regional and Local Landcare Coordinators*, NSW Landcare Gateway website. <https://landcare.nsw.gov.au/landcare-program/landcare-coordinators/>

Chapter 6

1 'The Beautiful South Coast: Trip to Eden', *Braidwood Dispatch and Mining Journal*, 17 April 1936, p. 2.

2 Donaldson S (2006) 'Stories about the Eurobodalla by Aboriginal People'. Environmental and Cultural Services, Moruya, NSW.

3 Donaldson S (2012) 'Djuuwin Women's Perspectives on the Moruya Deua River Catchment'. Southern Rivers Catchment Management Authority, Moruya, NSW.

4 NSW DPI (2016) 'Water Sharing Plan for the Deua River Unregulated and Alluvial Water Sources: Background Document'. NSW Department of Primary Industries, Sydney, NSW.

5 'Alternative lifestylers "in bushfire danger"', *The Canberra Times*, 11 December 1983, p. 6.

6 Green R (1985) 'Champion of Eurobodalla Ratepayers', *The Canberra Times*, 23 January, p. 18.

7 Lang R (1993) 'Potpourri of building materials key feature of Lisette's creative niche by the Deua River'. *The Canberra Times*, 31 December, p. 10.

8 'The house that Hutchings built', *The Canberra Times*, 10 August 1986, p. 4.

9 Mahoney D (1989) 'Tea for friends and travellers', *The Canberra Times*, 13 January, p. 4.

10 Fryer R (1989) 'Horsing around on the Deua', *The Canberra Times*, 7 June, p. 27.

11 Cartwright Z (2022) The highs and lows of life 'up the river', *About Regional*, 13 June (online). <https://aboutregional.com.au/the-highs-and-lows-of-life-up-the-river/398912/>

12 Australian Bureau of Statistics (2021) *Deua River Valley 2021 Census All persons QuickStats*, Australian Bureau of Statistics website, accessed 16 May 2025. <https://www.abs.gov.au/census/find-census-data/quickstats/2021/SAL11247>

13 Australian Bureau of Statistics (2021) *Australia 2021 Census All persons QuickStats*, Australian Bureau of Statistics website, accessed 16 May 2025. <https://www.abs.gov.au/census/find-census-data/quickstats/2021/AUS>

14 Eurobodalla Shire Council (1999) 'The Moruya River: A local perspective of change.' Eurobodalla Shire Council, Moruya, NSW, p. 12.

15 'Deua Rivercare', *Eurobodalla Newsletter*, May/June 2012, p. 8.

16 Bagavathiannan MV, Graham S, Ma Z, Barney JN, Coutts SR, Caicedo AL, De Clerk-Floate R, West NM, Blank L, Metcalf AL, Lacoste M, Moreno

CR, Evans JA, Burke I, Beckie H (2019) Considering weed management as a social dilemma bridges individual and collective interests. *Nature Plants* **5**, 343–351, p. 348.

[17] Ostrom E (1990) *Governing the Commons: The Evolution of Institutions for Collective Action*. Cambridge University Press, New York, USA, p. 90.

[18] Cox M, Arnold G, Tomás SV (2010) A review of design principles for community-based natural resource management. *Ecology and Society* **15**.

[19] Crow G, Allan G, Summers M (2002) Neither busybodies nor nobodies: Managing proximity and distance in neighbourly relations. *Sociology* **36**, 127–145.

[20] Crow G, Allan G, Summers M (2002) Neither busybodies nor nobodies: Managing proximity and distance in neighbourly relations. *Sociology* **36**, 127–145, p. 129.

[21] Bater F (2018) 'Weeds success is no myth', *Bay Post Moruya Examiner*, 5 April, n.p.

[22] Eurobodalla Landcare (2012–2017) *Momentum on the Deua River: Deua Rivercare group*, NSW Landcare Gateway website, accessed 16 May 2025 <https://landcare.nsw.gov.au/groups/eurobodalla-landcare-management-committee-elmc/momentum-on-the-deua-river/>.

[23] Naron F (2013) 'Koala habitat connected by Crossing students', *Narooma News*, 27 August, n.p.

[24] 'Peter Cormick reflects on Deua River bushfire', *Bay Post Moruya Examiner*, 20 May 2020, n.p.

[25] Australian Government Department of Health, Disability and Aging (n.d.) *Living with COVID-19 – COVID-19 ready kit*. Commonwealth of Australia, Canberra, accessed 16 May 2025. <https://www.health.gov.au/resources/publications/living-with-covid-19-covid-19-ready-kit?language=en>

[26] Explained by Trish Ellis in Donaldson, S. (2012) 'Djuuwin Women's Perspectives on the Moruya Deua River Catchment'. Southern Rivers Catchment Management Authority, Moruya, NSW, p. 49. Permission to include this quotation provided by AIATSIS.

[27] 'Courageous Women of the Deua River', *Braidwood Dispatch and Mining Journal*, 30 October 1942, p. 2.

Interlude: Bushfires and weeds go hand in hand

[1] Natural Hazards Research Australia (2023) 'Understanding the Black Summer bushfires through research: a summary of key findings from

the Bushfire and Natural Hazards CRC'. Natural Hazards Research Australia, Australia.

[2] Burgess T, Burgmann JR, Hall S, Holmes D, Turner E (2020) 'Black Summer: Australian newspaper reporting of the nation's worst bushfire season'. Monash Climate Change Communication Research Hub, Monash University, Melbourne, Victoria.

[3] Sommerville P (2019) Gippsland farmers are struggling through a green drought and say outsiders "don't understand"', *ABC News*, 19 July 2019. <https://www.abc.net.au/news/2019-07-19/gippsland-goes-through-green-drought/11322364>

[4] Brandt AJ, Png, GK, Jo I, McGrannachan C, Allen K, Peltzer DA, D'Antonio C, Dickie IA, French K, Leishman MR, Ostertag R, Parker IM, Stanley MC, Suding KN, Bellingham PJ (2023) Managing multi-species plant invasions when interactions influence their impact. *Frontiers in Ecology and the Environment* **21**, 370–379.

[5] Invasive Species Council (2014) 'Weeds and Climate Change', Invasive Species Council, Carlton, Victoria. <https://invasives.org.au/wp-content/uploads/2014/02/fs_weedsandclimatechange.pdf>

[6] Fenner Conference on the Environment (2020) 'Managing wild and weedy Australia across boundaries and disciplines: submission to the NSW Bushfire Inquiry'. University of Wollongong, Wollongong, NSW.

[7] 'Need for strategic weed response to bushfire crisis' [media release], The University of Wollongong, 28 February 2020 (online).

[8] 'Need for strategic weed response to bushfire crisis' [media release], The University of Wollongong, 28 February 2020 (online).

[9] Landcare Australia (2025) *Landcare Australia Bushfire Recovery Grants Program*, Landcare Australia website, accessed 30 May 2025. <https://landcareaustralia.org.au/project/landcare-australia-bushfire-recovery-grants-program/>

[10] Graham S (2012) Social relations and natural resource management: the significance of trust and power to solving a collective weed management problem. PhD thesis. Charles Sturt University, Wagga Wagga, NSW, Australia.

[11] Martin P (2020) 'Post fire plant identification – what to keep and what to remove' [video], *South East Local Land Services* and *Eurobodalla Shire Council*, YouTube, accessed 5 June 2025. <https://www.youtube.com/watch?v=Y3ul3UgCPC8&feature=youtu.be>

[12] BlazeAid is an Australian volunteer-based organisation that supports rural landholders after fires. After the 2019–2020 Black Summer bushfires, BlazeAid volunteers helped to rebuild fences, among other burnt infrastructure. BlazeAid Australia (2023) BlazeAid Australia website, accessed 30 May 2025. <https://blazeaid.com.au/>

Chapter 7

[1] NSW National Parks and Wildlife Service (2001) 'NSW Bitou Bush Strategy'. NSW National Parks and Wildlife Service, Sydney, NSW.

[2] Tathra Land Care (unpublished) 1993 Tathra Land Care. Video.

[3] Smith AR (1993) 'Bitou Bush – a noxious weed', Bega Valley Shire Council, Bega, NSW. Letter to householders.

[4] Friedel MH (2020) Unwelcome guests: a selective history of weed introductions to arid and semi-arid Australia. *Australian Journal of Botany* **68**, 75–99.

[5] Commonwealth Department of Commerce Cinema Branch (1935) 'The conquest of prickly pear' [video]. Commonwealth Prickly Pear Board, National Film and Sound Archives website, accessed 5 June 2025. <https://www.youtube.com/watch?v=T_K-m2rXWwc>

[6] Chapman D (1989) 'Coastal Dunes of New South Wales: Status and Management'. Soil Conservation Service of New South Wales, Sydney, NSW.

[7] Chapman D (1989) 'Coastal Dunes of New South Wales: Status and Management'. Soil Conservation Service of New South Wales, Sydney, NSW.

[8] Broese van Groenou EA, Wolfenden JAJ (2001) 'South Coast Regional Bitou Bush Strategy 2002'. Arnhem Environmental Impact Assessors, Armidale, NSW.

[9] Webber B, Scott J (22 May 2023) 'A botanical detective story: shedding light on the journey out of Africa for one of Australia's worst weeds.' *The Conversation*, accessed 2 December 2024. <https://theconversation.com/a-botanical-detective-story-shedding-light-on-the-journey-out-of-africa-for-one-of-australias-worst-weeds-205183>

[10] Chapman D (1989) 'Coastal Dunes of New South Wales: Status and Management'. Soil Conservation Service of New South Wales, Sydney, NSW, p. 100.

[11] Uhlmann A (1991) 'South Coast folk fight bitou bush'. *The Canberra Times*, 29 December, p. 2.

[12] Note: Chapman (1989) argues that it was identified as a suitable species in the 1940s.

[13] 'Beaches of Australia: A National Problem', *The Macleay Argus*, 27 January 1954, p. 3.

[14] McLeod K (1996) 'Bitou bush control in Tathra – pathway to success'. NSW Department of Land and Water Conservation, Bega, NSW.

[15] Ormonde P (1981) 'Beachwalk breaks new ground', *Tribune*, 11 March, p. 11.

[16] '18,000 demand sandmining halt', *Tribune*, 1 April 1981, p. 13.

[17] 'Sand stabilising bush causing concern', *The Canberra Times*, 3 August 1980, p. 6.

[18] Broese van Groenou EA, Wolfenden JAJ (2001) 'South Coast Regional Bitou Bush Strategy 2002'. Arnhem Environmental Impact Assessors, Armidale, NSW.

[19] 'Shire council poll nominations open: Minister says "no ridings"'. *The Canberra Times*, 30 October 1983, p. 6.

[20] 'Groundcover "a menace" despite its beauty', *The Canberra Times*, 23 September 1984, p. 7.

[21] Broese van Groenou EA, Wolfenden JAJ (2001) 'South Coast Regional Bitou Bush Strategy 2002'. Arnhem Environmental Impact Assessors, Armidale, NSW, p. 6.

[22] McLeod K (1996) 'Bitou bush control in Tathra – pathway to success'. NSW Department of Land and Water Conservation, Bega, NSW.

[23] 'Move to declare bitou bush a noxious plant', *The Canberra Times*, 1 November 1985, p. 12.

[24] Bega Environment Network (1993) 'Environment Network Jobskills Program. Bitou bush control program – Tathra'. Bega Environment Network, Bega, NSW.

[25] 'Bitou takes over Tathra', *Bega District News*, 2 April 1993.

[26] Love A (1984) Distribution of bitou bush along the New South Wales Coast. In *Bitou bush and Boneseed: Proceedings of a conference on Chrysanthemoides monilifera*. 8–9 August, Port Macquarie. (Eds. Love A, Dyason R) p. 53–64. NSW National Parks and Wildlife Service and Department of Agriculture NSW, Port Macquarie, NSW.

[27] Kelly J (unpublished) '1993 Tathra Land Care Notebook'. Tathra, NSW.

[28] 'News flash: Bitou bush takes over Tathra', flyer advertising a public meeting in 1993. Emphasis added.

[29] 'Bitou bush meeting success', *Bega District News*, 23 April 1993.

[30] Kelly J (unpublished) Notebook.

[31] Smith AR (1993) 'Bitou Bush – a noxious weed', Bega Valley Shire Council, Bega, NSW. Letter to householders, emphasis added.

[32] Holtkamp RH (1993) 'Bitou bush and boneseed bulletin'. NSW Department of Agriculture, Tamworth, NSW.

[33] 'Jobskills Landcare project', *ENet News: Newsletter of the Environment Network*, October 1993, p. 1.

[34] McLeod K (1996) 'Bitou bush control in Tathra – pathway to success'. NSW Department of Land and Water Conservation, Bega, NSW, p. 10.

[35] Bega Valley Shire Council (2017) '2016–17 Annual Report'. Bega Valley Shire Council, Bega, NSW.

[36] Winkler MA, Cherry H, Downey PO (2008) 'Bitou Bush Management Manual: current management and control options for bitou bush (*Chrysanthemoides monilifera* ssp. *rotundata*) in Australia'. NSW Department of Environment and Climate Change, Sydney, NSW.

[37] McLeod K (1996) 'Bitou bush control in Tathra – pathway to success'. NSW Department of Land and Water Conservation, Bega, NSW.

[38] NSW National Parks and Wildlife Service (2001) 'NSW Bitou Bush Strategy', NSW National Parks and Wildlife Service, Sydney, NSW.

[39] Broese van Groenou EA and Wolfenden JAJ (2002) 'South Coast Regional Bitou Bush Strategy', Arnhem Environmental Impact Assessors, Armidale, NSW.

Chapter 8

[1] Candelo Historical Committee (1985) *Candelo Recollections*. Candelo Historical Committee, Candelo, NSW.

[2] Blain G (1999) *Candelo*. Penguin, Ringwood, Victoria. Emphasis in original.

[3] Others explain that the town was named after a village in Italy. Peter Imlay, a 'free settler', arrived in Australia in about 1820 and established himself in the Bega district. He later built a house – Candelo House – on land he occupied on the western side of the creek some 30 years before the town was formally established. For more information about the Imlays see: Wellings HP (1930) 'The Brothers Imlay: Pioneers of the Eden District', *Sydney Morning Herald*, 28 June 1930, p. 11. Candelo Historical Committee (1985) *Candelo Recollections*. Candelo Historical Committee, Candelo, NSW.

[4] Dwyer J (2017) Willows and the Australian Landscape. *Australian Garden History* **29**, 20–22.

[5] Dwyer J (2017) Willows and the Australian Landscape. *Australian Garden History* **29**, 20–22.

[6] 'Local and general news', *Cobargo Chronical*, 24 May 1919, p. 2.

[7] Sargeant H (1952) Garden Trees and Shrubs in Australasia. Colorgravure, Melbourne, Victoria.

[8] Elg H (2019) Watch how Australia's famous cricket bats are made near Daylesford, *Bendigo Advertiser*, 22 February.

[9] Dwyer J (2017) Willows and the Australian Landscape. *Australian Garden History* **29**, 20–22.

[10] Cremer K, Van Kraayenoord C, Parker N, Streatfield S (1995) Willows spreading by seed – implications for Australian river management. *Australian Journal of Soil and Water Conservation* **8**, 18–27.

[11] Jamieson T (1998) 'Time to Weed the Unwelcome Willow'. *Sydney Morning Herald*, 12 February, p. 5.

[12] Ladson A, Gerrish G, Carr G, Thexton E (1997) 'Willows along Victorian waterways: towards a willow management strategy'. Victorian Department of Natural Resources and Environment, Melbourne, Victoria.

[13] Holland Clift S (n.d.) 'Weeds of National Significance. Willows Identification: An Essential Skill for Successful Willow Management. Willow Resource Sheet 2'. Victorian Department of Primary Industries Victoria, Melbourne, Victoria.

[14] Ladson A, Gerrish G, Carr G, Thexton E (1997) 'Willows along Victorian waterways: towards a willow management strategy'. Victorian Department of Natural Resources and Environment, Melbourne, Victoria.

[15] Barlow G (2001) No weeping for the willow. *The Weekly Times*, 24 January, p. 9.

[16] Rutherford I (2018) 'Why is everyone talking about natural sequence farming?' *The Conversation*, accessed 5 June 2025. <https://theconversation.com/why-is-everyone-talking-about-natural-sequence-farming-106232>

[17] Callow N, Bell RA (2021) 'The applicability, efficacy and risks of natural sequence farming in the dryland agricultural zone of south west Western Australia'. University of Western Australia, Perth, WA.

[18] Dwyer J (2017) Willows and the Australian Landscape. *Australian Garden History* **29**, 20–22, p. 22.

[19] Simpson I (2022) *Willowman*. Hachette Australia, Sydney, NSW.

[20] 'A Tour to the South: The Bega District', *Empire*, 20 Nov 1871, p. 4.

[21] The Scout (1887) 'Some dairying notes', *Sydney Mail*, 30 April, p. 904.

[22] 'Candelo', *Southern Star*, 28 August 1907, p. 4.

[23] 'Candelo', *Two Fold Bay Magnet*, 16 August 1909, p. 8.

[24] 'News and Comment', *Southern Record and Advertiser*, 15 November 1919, p. 4.

[25] 'Local and general news', *Cobargo Chronical*, 25 August 1923, p. 2.

[26] 'News and Comment', *Southern Record and Advertiser*, 15 November 1919, p. 4.

[27] 'The Southern Star', *Southern Star*, 18 April 1923, p. 2.

[28] 'Notification of granting of special leases', *Government Gazette of the State of NSW*, 28 August 1924, p. 4177.

[29] 'Imlay Shire Council', *Bega District News*, 30 March 1939, p. 3.

[30] 'Local and general', *Southern Record and Advertiser*, 21 October 1916, p. 1.

[31] 'Wistful whispers', *Southern Record and Advertiser*, 29 September 1917, p. 4.

[32] 'Imlay Shire Council', *Southern Record and Advertiser*, 25 January 1919, p. 1.

[33] 'Local and general', *The Southern Record and Advertiser*, 29 January 1932, p. 1.

[34] 'Mumbulla Shire', *Bega District News*, 3 October 1932, p. 2.

[35] 'Should willows be planted in Bega River?', *Bega District News*, 7 November 1932, p. 2.

[36] Lester Firth and Associates (1989) 'Candelo: Heritage Study'. Lester Firth Associates Pty Ltd, Canberra, ACT.

[37] Jamieson T (1998) 'Time to Weed the Unwelcome Willow'. *Sydney Morning Herald*, 12 February 1998, p. 5.

[38] Nairn G (2003) $292,081 Fed Govt priority drought and fire funding, accessed 24 February 2025. <https://parlinfo.aph.gov.au/parlInfo/download/media/pressrel/ABDA6/upload_binary/abda62.pdf;fileType=application%2Fpdf#search=%22candelo%20creek%22>

[39] Dion D (2023) 'How one woman helped to reveal a river', *Bega District News*, Bega, 27 January, p. 4.

[40] 'Stage Two for Candelo Landcare', *Bega District News*, 7 May 2004.

[41] Koori, or Koorie, is a word used to describe Aboriginal people in Victoria and southern New South Wales. DeadlyStory (n.d.) *I love being a Koori kid because…* Deadly Story website, accessed 30 May 2025. <https://deadlystory.com/page/culture/our-stories/i-love-being-a-koorie-kid-because>

[42] 'Kooris are tops', *Bega District News*, 6 January 2004.

[43] 'Stage Two for Candelo Landcare', *Bega District News*, 7 May 2004.

[44] Nairn G (2006) Investing in our local environment. *Summit to Surf: Gairy Nairn MP*, Canberra, ACT.

[45] McDonald A (2018) Tireless Landcare anti-erosion project turned to mulch. *Bega District News*, 1 June, n.p.

[46] Bega Valley Shire Council (2015) 'Bega Valley Annual Report 2014–2015'. Bega Valley Shire Council, Bega, NSW.

[47] 'Flood recovery help', *Bega District News*, 31 May 2011, n.p.

[48] 'Making a difference on Clean Up Day', *Bega District News*, 12 March 2004, p. 12.

[49] 'Trees, trees trees!', *Candelo Connect*, 19 August 2009, p. 8.

[50] 'Candelo Public School', *Bega District News*, 13 June 2006, p. 1.

[51] Graham S, McGinness H, O'Connell D (2008) Climatic Drivers of Revegetation Management Practices in Australia: Analysis of a Social Survey. *Small-Scale Forestry* **7**, 183–203.

[52] Bettanin J (2022) Candelo Community Celebrates Caring for the Creek. NSW Landcare Gateway website, accessed 19 February 2025. <https://landcare.nsw.gov.au/groups/far-south-coast-landcare-association-inc/candelo-community-celebrates-caring-for-the-creek-3>

[53] Smyth B (2014) 'Art project reimagines Bega Cheese waste', *Bega District News*, 14 July 2014.

[54] Department of Climate Change, Energy, the Environment and Water (2022) Accessed 20 May 2025. *Planting Trees for The Queen's Jubilee Program: Approved Projects*, Commonwealth Department of Climate Change, Energy, the Environment and Water, Canberra. <https://www.dcceew.gov.au/sites/default/files/documents/queens-jubilee-funded-project-list.pdf>

[55] Dion D (2023) 'How one woman helped to reveal a river', *Bega District News*, Bega, 27 January, p. 4.

Chapter 9

[1] Serrated Tussock (2024) *Thompson's Legacy – serrated tussock in the Ladevale District*, accessed 25 March 2024. <http://www.serratedtussock.com.au/?i=361&thompsons-legacy-serrated-tussock-in-the-ladevale-district>

[2] Graham S (2012) Social relations and natural resource management: the significance of trust and power to solving a collective weed management problem. PhD thesis. Charles Sturt University, Wagga Wagga, NSW, p. 156.

[3] Agriculture Victoria (2020) *Serrated Tussock*. Agriculture Victoria website, accessed 15 January 2025. <https://agriculture.vic.gov.au/biosecurity/weeds/weeds-information/serrated-tussock>

[4] Victorian Serrated Tussock Working Party (2018) 'Best Practice Management Guide: Serrated Tussock'. Agriculture Victoria, Melbourne, Victoria.

[5] Serrated Tussock (2024) *Thompson's Legacy – serrated tussock in the Ladevale District,* accessed 25 March 2024. <http://www.serratedtussock.com.au/?i=361&thompsons-legacy-serrated-tussock-in-the-ladevale-district>

[6] 'Weeds more than a wee problem', *Sunbury/Macedon Ranges Leader,* 12 December 2006.

[7] Osmond R, Veebeek M, McLaren DA, Michelmore M, Wicks B, Grech CJ, Fullerton P (2008) *Serrated tussock – National best practice manual.* Victorian Department of Primary Industries, Melbourne, Victoria.

[8] NSW Department of Primary Industries (2025) *Serrated tussock (Nassella trichotoma).* NSW WeedWise website, accessed 5 June 2025. <https://weeds.dpi.nsw.gov.au/Weeds/Details/123>

[9] Goninon C (1998) Serrated tussock in Tasmania. *Plant Protection Quarterly* **13**, 98–99.

[10] Humphries T, Dowling K, Turville C, Sinclair S, Florentine S (2020) Ecology, distribution and control of the invasive weed Nassella trichotoma (Nees) Hack. ex Arechav.: A global review of current and future challenges. *Weed Research* **60**, 392–405.

[11] Victorian Serrated Tussock Working Party (2025) *Impacts of Serrated Tussock,* accessed 15 January 2025. <https://serratedtussock.com/about/impacts-of-serrated-tussock>

[12] Ayers L, Leech F (2010) Serrated tussock (Nassella trichotoma) Identification and control. In NSW Department of Primary Industries (Ed.), Primefact Vol. 44, NSW Department of Industry and Investment.

[13] 'Helicopter to spray serrated tussock', *The Canberra Times*, 11 September 1980, p. 16.

[14] Graham S, Wyllie M, Wilkerson M, Williams M, Sharp A, Cherry H, Martin P, Campbell R, Hawkes G (2022) Measuring the success of cross-tenure collaborative weed management: insights codeveloped with practitioners. *Invasive Plant Science and Management* **15**, 183–193.

[15] Marshall GR, Coleman MJ, Sindel BM, Reeve IJ, Berney PJ (2016) Collective action in invasive species control, and prospects for community-based governance: The case of serrated tussock (Nassella trichotoma) in New South Wales, Australia. *Land Use Policy* **56**, 100–111.

[16] Graham S, Rogers S (2017) How local landholder groups collectively manage weeds in south-eastern Australia. *Environmental Management* **60**, 396–408.

[17] Eden Local Aboriginal Land Council (2025) *Route*, the Bundian Way website, accessed 15 January 2025. <https://www.bundianway.com.au/route>

[18] Like many places in Australia, the Bundian Way is burdened with stories of massacres, dispossession and disenfranchising of Aboriginal peoples. See for example Baker M, Cutcher L, Ormiston J (2023) The Bundian way: An Indigenous-led cross-sector partnership in place through time. *Journal of Business Ethics* **184**, 877–894.

[19] Australian Bureau of Statistics (2021) *Towamba 2021 Census All person QuickStats*. Australian Bureau of Statistics website, accessed 16 May 2025. <https://www.abs.gov.au/census/find-census-data/quickstats/2021/SAL13945>.

[20] Australian Bureau of Statistics (2021) *Burragate 2006 Census All person QuickStats*. Australian Bureau of Statistics website, accessed 20 May 2025. <https://www.abs.gov.au/census/find-census-data/quickstats/2021/SAL10700>

[21] Australian Bureau of Statistics (2021) *Pericoe 2021 Census All person QuickStats*. Australian Bureau of Statistics website, accessed 20 May 2025. <https://www.abs.gov.au/census/find-census-data/quickstats/2021/SAL13198>

[22] 'Towamba Public School turns 150', *Bombala Times*, 31 July 2012.

[23] Australian Bureau of Statistics (2006) *Towamba 2006 Census All person QuickStats*. Australian Bureau of Statistics website, accessed 16 May 2025. <https://www.abs.gov.au/census/find-census-data/quickstats/2006/SSC19095>.

[24] Australian Bureau of Statistics (2021) *Towamba 2021 Census All person QuickStats*. Australian Bureau of Statistics website, accessed 16 May 2025. <https://www.abs.gov.au/census/find-census-data/quickstats/2021/SAL13945>.

[25] Herbert A (2013) Towamba Landcare – Community Leaders. A community driving land at the catchment level. In *Proceedings of the 17th NSW Weeds*

Conference. 9–12 September, Corowa. The Weeds Society of NSW, Corowa, New South Wales.

[26] 'Slowing spread of serrated tussock', *Bega District News*, 17 October 2011.

[27] Herbert A (2013) Towamba Landcare – Community Leaders. A community driving land at the catchment level. In *Proceedings of the 17th NSW Weeds Conference*. 9–12 September, Corowa. The Weeds Society of NSW, Corowa, New South Wales.

[28] Herbert A (2013) Towamba Landcare – Community Leaders. A community driving land at the catchment level. In *Proceedings of the 17th NSW Weeds Conference*. 9–12 September, Corowa. The Weeds Society of NSW, Corowa, New South Wales.

[29] Graham S (2013) Three cooperative pathways to solving a collective weed management problem. *Australasian Journal of Environmental Management* **20**, 116–129.

[30] Herbert A (2013) Towamba Landcare – Community Leaders. A community driving land at the catchment level. In *Proceedings of the 17th NSW Weeds Conference*. 9–12 September, Corowa. The Weeds Society of NSW, Corowa, New South Wales.

[31] Lewis D (2007) 'Volunteers clean up Love Grass in Rocky Hall / New Buildings', *Wyndham Whisper*, July–August, p. 1.

[32] Koori, or Koorie, is a word used to describe Aboriginal people in Victoria and southern New South Wales. DeadlyStory (n.d.) *I love being a Koori kid because…* Deadly Story website, accessed 30 May 2025. <https://deadlystory.com/page/culture/our-stories/i-love-being-a-koorie-kid-because>

[33] Cavanagh M (2014) Landcare cuts will knock the wider community, *ABC News*, 2 June, n.p.

[34] Lewis D (2017) 'Towamba Valley Landcare Group News', *Wyndham Whisper*, September–October, p. 3.

[35] Graham S, Rogers S (2017) How local landholder groups collectively manage weeds in south-eastern Australia. *Environmental Management* **60**, 396–408.

[36] Graham S, Rogers S (2017) How local landholder groups collectively manage weeds in south-eastern Australia. *Environmental Management* **60**, 396–408, p. 401.

[37] Lewis D (2007) 'Volunteers clean up Love Grass in Rocky Hall / New Buildings', *Wyndham Whisper*, July–August, p. 1.

[38] 'Landcare News – November 2007', *Wyndham Whisper*, November–December 2007, p. 2.

[39] Lewis D (2017) 'Towamba Valley Landcare Group News', *Wyndham Whisper*, September–October, p. 3.

[40] Bryant S, Cavanagh M, Condon M (2014) 'Pasture cropping pioneer Col Seis receives Bob Hawke Landcare Award', *ABC News*, 19 September, n.p.

[41] 'How to identify and control feral pig population', *Yass Tribune*, 4 December 2014, n.p.

[42] Lewis D (2019) 'Wild dogs in the Towamba Valley', *Wyndham Whisper*, May, p. 6.

[43] Lewis D (2017) 'Towamba Valley Landcare Group News', *Wyndham Whisper*, September–October, p. 3.

[44] Graham S, Rogers S (2017) How local landholder groups collectively manage weeds in south-eastern Australia. *Environmental Management* **60**, 396–408.

Interlude: Beyond volunteer enthusiasts

[1] Australian Bureau of Statistics (2021) *Australia 2021 Census All persons QuickStats*, Australian Bureau of Statistics website, accessed 16 May 2025. < https://www.abs.gov.au/census/find-census-data/quickstats/2021/AUS>

[2] Biddle N, Boyer C, Gray M, Jahromi M (2022) 'Volunteering in Australia: The Volunteer Perspective.' Volunteering Australia, accessed 3 February 2025. <https://volunteeringhub.org.au/wp-content/uploads/2024/06/Volunteering-in-Australia-2022-The-Volunteer-Perspective.pdf>

[3] The National Council for Voluntary Organisations (2023) 'UK Civil Society Almanac 2023. Data. Trends. Insights.' The National Council for Voluntary Organisations, London, United Kingdom, accessed 3 February 2025. <https://www.ncvo.org.uk/news-and-insights/news-index/uk-civil-society-almanac-2023/volunteering>

[4] Australian Bureau of Statistics (2011) *Australia 2011 Census All persons QuickStats*, Australian Bureau of Statistics website, accessed 16 May 2025. <https://www.abs.gov.au/census/find-census-data/quickstats/2011/0>

[5] Australian Bureau of Statistics (2021) *Australia 2021 Census All persons QuickStats*, Australian Bureau of Statistics website, accessed 16 May 2025. <https://www.abs.gov.au/census/find-census-data/quickstats/2021/AUS>

[6] For example, Biddle N, Boyer C, Gray M, Jahromi M (2022) 'Volunteering in Australia: The Volunteer Perspective.' Volunteering Australia, accessed 3 February 2025. <https://volunteeringhub.org.au/wp-content/uploads/2024/06/Volunteering-in-Australia-2022-The-Volunteer-Perspective.pdf>

[7] Graham S, Barnett J, Fincher R, Hurlimann A, Mortreux C (2014) Local values for fairer adaptation to sea-level rise: A typology of residents and their lived values in Lakes Entrance, Australia. *Global Environmental Change* **29**, 41–52.

[8] Graham S, Barnett J, Mortreux C, Hurlimann A, Fincher R (2018) Local values and fairness in climate change adaptation: insights from marginal rural Australian communities. *World Development* **108**, 332–343.

[9] McDougle LM, Greenspan I, Handy F (2011) Generation green: understanding the motivations and mechanisms influencing young adults' environmental volunteering. *International Journal of Nonprofit and Voluntary Sector Marketing* **16**, 325–341.

[10] Biddle N, Boyer C, Gray M, Jahromi M (2022) 'Volunteering in Australia: The Volunteer Perspective.' Volunteering Australia, accessed 3 February 2025. <https://volunteeringhub.org.au/wp-content/uploads/2024/06/Volunteering-in-Australia-2022-The-Volunteer-Perspective.pdf>

[11] Willems J, Dury S (2017) Reasons for not volunteering: overcoming boundaries to attract volunteers. *The Service Industries Journal* **37**, 726–745.

[12] Biddle N, Boyer C, Gray M, Jahromi M (2022) 'Volunteering in Australia: The Volunteer Perspective.' Volunteering Australia, accessed 3 February 2025. <https://volunteeringhub.org.au/wp-content/uploads/2024/06/Volunteering-in-Australia-2022-The-Volunteer-Perspective.pdf>

[13] Willems J, Dury S (2017) Reasons for not volunteering: overcoming boundaries to attract volunteers. *The Service Industries Journal* **37**, 726–745.

[14] Biddle N, Boyer C, Gray M, Jahromi M (2022) 'Volunteering in Australia: The Volunteer Perspective.' Volunteering Australia, accessed 3 February 2025. <https://volunteeringhub.org.au/wp-content/uploads/2024/06/Volunteering-in-Australia-2022-The-Volunteer-Perspective.pdf>

[15] Willems J, Dury S (2017) Reasons for not volunteering: overcoming boundaries to attract volunteers. *The Service Industries Journal* **37**, 726–745.

[16] Biddle N, Boyer C, Gray M, Jahromi M (2022) 'Volunteering in Australia: The Volunteer Perspective.' Volunteering Australia, accessed 3

February 2025. <https://volunteeringhub.org.au/wp-content/uploads/2024/06/Volunteering-in-Australia-2022-The-Volunteer-Perspective.pdf>

[17] Walsh A, Kingwell R (2021) Economic implications of the loss of glyphosate and paraquat on Australian mixed enterprise farms. *Agricultural Systems* **193**, 103207.

[18] Local Land Services (n.d.) NSW Landcare Enabling Program 2023–2027. NSW Local Land Services webpage, accessed 30 May 2025. <https://www.lls.nsw.gov.au/what-we-do/our-major-projects/nsw-landcare-program>

Chapter 10

[1] Such as the introduction of the notion of 'shared responsibility' with the *NSW Biosecurity Act 2015*.

[2] Sobels J, Curtis A, Lockie S (2001) The role of Landcare group networks in rural Australia: exploring the contribution of social capital. *Journal of Rural Studies* **17**, 265–276.

[3] Webb T, Cary J (2005) Social capital and natural resource management: An application to Landcare. *Rural Society* **15**, 119–131.

[4] Compton E, Beeton RB (2012) An accidental outcome: Social capital and its implications for Landcare and the 'status quo'. *Journal of Rural Studies* **28**, 149–160.

[5] Minato W, Curtis A, Allan C (2010) Social norms and natural resource management in a changing rural community. *Journal of Environmental Policy & Planning* **12**, 381–403.

[6] Quinn D (2011) Park under Fire. *Wildlife Australia Magazine* **Autumn**, 34–37.

[7] Berry ZC, Wevill K, Curran TJ (2011) The invasive weed Lantana camara increases fire risk in dry rainforest by altering fuel beds. *Weed Research* **51**, 525–533.

[8] Putnam R (2000) *Bowling Alone: The collapse and revival of American Community*. Simon & Schuster Paperbacks, New York, USA.

[9] Allen J (2022) *Find Your People: Building Deep Community in a Lonely World.* WaterBrook, New York, USA.

[10] Elinor Ostrom identified eight design principles of long-enduring common pool resource institutions. The three that align with the attributes discussed in this chapter are: 1) clearly defined boundaries; 2) conflict-resolution mechanisms; and 3) nested enterprises. Ostrom E

(1990) *Governing the Commons: The Evolution of Institutions for Collective Action*. Cambridge University Press, New York, USA, p. 90.

[11] Muthu Bagavathiannan and colleagues identified four additional design principles that are relevant to weed-management collective action. The three that align with the attributes in this chapter are: 1) contributors agree to a shared goal; 2) contributions are made transparently; and 3) strong working relationships or shared values among contributors. Bagavathiannan MV, Graham S, Ma Z, Barney JN, Coutts SR, Caicedo AL, De Clerk-Floate R, West NM, Blank L, Metcalf AL, Lacoste M, Moreno CR, Evans JA, Burke I, Beckie H (2019) Considering weed management as a social dilemma bridges individual and collective interests. *Nature Plants* **5**, 343–351, p. 348.

[12] Bagavathiannan MV, Graham S, Ma Z, Barney JN, Coutts SR, Caicedo AL, De Clerk-Floate R, West NM, Blank L, Metcalf AL, Lacoste M, Moreno CR, Evans JA, Burke I, Beckie H (2019) Considering weed management as a social dilemma bridges individual and collective interests. *Nature Plants* **5**, 343–351.

[13] Ostrom E (1990) *Governing the Commons: The Evolution of Institutions for Collective Action*. Cambridge University Press, New York, USA.

[14] Graham S, Metcalf AL, Gill N, Niemiec R, Moreno C, Bach T, Ikutegbe V, Hallstrom L, Ma Z, Lubeck A (2019) Opportunities for better use of collective action theory in research and governance for invasive species management. *Conservation Biology* **33**, 275–287.

[15] Wilson DS, Ostrom E, Cox ME (2013) Generalizing the core design principles for the efficacy of groups. *Journal of Economic Behavior & Organization* **90**, S21–S32.

[16] Figuera SG., Babcock, B., Lubell, M., & McRoberts, N. (2022). Collective action in the area-wide management of an invasive plant disease. *Ecology and Society*, *27*(2).

[17] Sullivan A, York AM (2021) Collective action for changing forests: A spatial, social-ecological approach to assessing participation in invasive plant management. *Global Environmental Change* **71**, 102366.

[18] Cox M, Arnold G, Tomás SV (2010) A review of design principles for community-based natural resource management. *Ecology and Society*, **15**.

[19] Bagavathiannan MV, Graham S, Ma Z, Barney JN, Coutts SR, Caicedo AL, De Clerk-Floate R, West NM, Blank L, Metcalf AL, Lacoste M, Moreno CR, Evans JA, Burke I, Beckie H (2019) Considering weed management

as a social dilemma bridges individual and collective interests. *Nature Plants* **5**, 343–351.

[20] Paull M, Redmond J (2011) Succession in community organisations: Newcomers and the 'purple circle'. *Third Sector Review* **17**, 131–152.

[21] Graham S, Rogers S (2017) How local landholder groups collectively manage weeds in south-eastern Australia. *Environmental Management* **60**, 396–408.

[22] Compton, E., & Beeton, R. B. (2012). An accidental outcome: Social capital and its implications for Landcare and the 'status quo'. *Journal of Rural Studies* **28**, 149–160.

[23] Glowacki L, von Rueden C (2015) Leadership solves collective action problems in small-scale societies. *Philosophical Transactions of the Royal Society B: Biological Sciences* **370**, 20150010.

[24] Cox M, Arnold G, Tomás SV (2010) A review of design principles for community-based natural resource management. *Ecology and Society*, **15**.

[25] Ruttan L (2006) Sociocultural heterogeneity and the commons. *Current Anthropology*, *47*(5), 843–853.

[26] Ostrom E (1990) *Governing the Commons: The Evolution of Institutions for Collective Action*. Cambridge University Press, New York, USA.

[27] The view of conflict as productive resonates with academic research on Agonistic Planning Theory, which sees conflict as a productive force for social change. See for example Kühn M (2021) Agonistic planning theory revisited: The planner's role in dealing with conflict. *Planning Theory* **20**, 143–156.

[28] Cox M, Arnold G, Tomás SV (2010) A review of design principles for community-based natural resource management. *Ecology and Society*, **15**.

[29] Campbell A (unpublished) Landcare: Grassroots environmental governance for the 21st century. South East Landcare Muster keynote, 29 May 2025, Bungendore.

[30] Landcare Australia (n.d.) Find a group on the National Landcare Directory, Landcare Australia website, accessed 30 May 2025. <https://landcareaustralia.org.au/find-a-group/>

[31] In NSW, Landcare coordinators are listed on the NSW Landcare Gateway website: https://landcare.nsw.gov.au/landcare-program/landcare-coordinators/

INDEX

www.ingramcontent.com/pod-product-compliance
Lightning Source LLC
LaVergne TN
LVHW090315210926
848480LV00001B/251